Springer Theses

Recognizing Outstanding Ph.D. Research

Aims and Scope

The series "Springer Theses" brings together a selection of the very best Ph.D. theses from around the world and across the physical sciences. Nominated and endorsed by two recognized specialists, each published volume has been selected for its scientific excellence and the high impact of its contents for the pertinent field of research. For greater accessibility to non-specialists, the published versions include an extended introduction, as well as a foreword by the student's supervisor explaining the special relevance of the work for the field. As a whole, the series will provide a valuable resource both for newcomers to the research fields described, and for other scientists seeking detailed background information on special questions. Finally, it provides an accredited documentation of the valuable contributions made by today's younger generation of scientists.

Theses are accepted into the series by invited nomination only and must fulfill all of the following criteria

- They must be written in good English.
- The topic should fall within the confines of Chemistry, Physics, Earth Sciences, Engineering and related interdisciplinary fields such as Materials, Nanoscience, Chemical Engineering, Complex Systems and Biophysics.
- The work reported in the thesis must represent a significant scientific advance.
- If the thesis includes previously published material, permission to reproduce this must be gained from the respective copyright holder.
- They must have been examined and passed during the 12 months prior to nomination.
- Each thesis should include a foreword by the supervisor outlining the significance of its content.
- The theses should have a clearly defined structure including an introduction accessible to scientists not expert in that particular field.

More information about this series at http://www.springer.com/series/8790

Philip A. Thomas

Narrow Plasmon Resonances in Hybrid Systems

Doctoral Thesis accepted by
the University of Manchester, Manchester, UK

Author
Dr. Philip A. Thomas
Department of Physics and Astronomy
University of Exeter
Exeter, UK

Supervisor
Prof. Alexander N. Grigorenko
School of Physics and Astronomy
University of Manchester
Manchester, UK

ISSN 2190-5053 ISSN 2190-5061 (electronic)
Springer Theses
ISBN 978-3-030-07367-1 ISBN 978-3-319-97526-9 (eBook)
https://doi.org/10.1007/978-3-319-97526-9

This Springer imprint is published by the registered company Springer Nature Switzerland AG
The registered company address is: Gewerbestrasse 11, 6330 Cham, Switzerland

Their minds sang with the ecstatic knowledge that either what they were doing was completely and utterly and totally impossible or that physics had a lot of catching up to do.

Douglas Adams

Supervisor's Foreword

In the last two decades, the field of plasmonics has attracted a lot of attention. Plasmonics is devoted to studies of the interaction of light with subwavelength objects that contain free-electron plasma. There are two main reasons for the increased interest in plasmonics. First, developments in nanofabrication have allowed the fabrication of various artificial materials which are capable of carrying optical plasmons. These artificial materials formed the basis for a new experimental field of investigation (optical metamaterials) promising a whole set of new phenomena: light compression into subwavelength regions, electromagnetic field enhancement, perfect lensing, negative refraction, topological darkness, etc. Second, being the last relatively unexplored branch of optics, plasmonics carries the torch of optics which is always set not only to unveil exciting new science but also to deliver important applications useful for society and the general public. Indeed, optics was the first useful physical discipline (the laws of light refraction and reflection, fibre optics and lenses without distortion were established as early as the tenth century AD) and has continued to bring forth a variety of useful tools and installations: glasses, lenses, polarizers, lasers, interferometers, optical fibres, nonlinear frequency doublers, etc. Plasmonics has continued the long-standing tradition of optics and delivered applications in the areas of ultrasensitive biosensing, materials with the designed optical response, nonlinear optics, etc.

The efforts of many groups and advances in our understanding of the interaction of light with subwavelength plasmonic materials have allowed unique optical responses to be tailored at the nanoscale. In particular, it was recently found that metal nanostructures (capable of supporting surface plasmons) can be designed to possess spectrally narrow plasmon resonances, which are of particular interest due to their exceptional sensitivity to their local environment. Combining plasmonic nanostructures with other materials in hybrid systems enabled this sensitivity to be exploited in a broad range of applications. In this thesis, Philip Thomas explores two different approaches to attaining narrow plasmon resonances: in gold nanoparticle arrays by utilising diffraction coupling and in copper thin films covered by a protective graphene layer. The performances of these resonances were evaluated for a number of applications. For example, nanoparticle arrays along with an atomic heterostructure were used as elements in a nanomechanical electro-optical modulator, capable of strong, broadband light modulation. Strong coupling between diffraction-coupled plasmon resonances in a gold nanoparticle array and guided modes in a dielectric slab was applied to construct a hybrid waveguide. The extreme phase sensitivity of graphene-protected copper was utilised to detect trace quantities of small toxins in solution far below the detection limit of commercial surface plasmon resonance sensors. One can expect that Philip Thomas' thesis is the first real step towards combining the extraordinary properties of plasmonics and 2D materials into smart hybrid optical devices, promising the realisation of valuable new applications.

Manchester, UK
April 2018

Prof. Alexander N. Grigorenko

Abstract

Surface plasmons are collective oscillations of free electrons excited at a metal–dielectric interface by incident light. They possess a broad set of interesting properties including a high degree of tunability, the generation of strong field enhancements close to the metal's surface and high sensitivity to their adjacent dielectric environment.

It is possible to enhance the sensitivity of plasmonic systems by using narrow plasmon resonances. In this thesis, two approaches to narrowing surface plasmon resonances have been studied: diffraction coupling of localised surface plasmon resonances in gold nanoarrays and the use of graphene-protected copper thin films. Applications of these approaches in hybrid systems have been considered for modulation, waveguiding and biosensing.

Arrays of gold nanostripes fabricated on a gold sublayer have been used to create extremely narrow plasmon resonances using diffraction coupling of localised plasmon resonances with quality factors up to a value of $Q \sim 300$, among the highest reported in the literature. The nanostructures were designed to give the narrowest resonance at the telecommunication wavelength of 1.5 μm, allowing for this array geometry to be used in hybrid systems for proof-of-concept optoelectronic devices.

The gold nanostripe array was used in a hybrid nanomechanical electro-optical modulator along with hexagonal boron nitride (hBN) and graphene. The modulator was fabricated with an air gap between the nanoarray and the hexagonal boron nitride/graphene. Applying a gate voltage across the device moves the hBN towards the nanoarray, resulting in broadband modulation effects from the ultraviolet through to the mid-infrared dependant on the motion of the hBN instead of graphene gating.

The deposition of a 400-nm hafnium(IV) oxide film on top of the gold nanoarray created a structure capable of guiding modes at 1.5 μm. The hybrid air–dielectric–stripe waveguide is capable of guiding modes over a distance of 250 μm.

Copper thin films have stronger plasmon resonances and higher phase sensitivity than gold thin films. Transferring a graphene sheet on the copper prevents oxidation of the copper. A feasibility study of this hybrid system has shown that phase-sensitive graphene-protected copper biosensing can detect HT-2 mycotoxin with over four orders of magnitude greater sensitivity than commercially available gold-based surface plasmon resonance biosensing systems.

In summary, two methods of attaining narrow plasmon resonances have been demonstrated and their promise in modulation, waveguiding and biosensing has been demonstrated.

Acknowledgements

This thesis is the culmination of a series of adventures and near-misadventures. The successes have far outnumbered the failures during this Ph.D., something that can only be attributed to the kindness and generosity shown to me over the past few years by a large number of people.

I was lucky to start my Ph.D. when Ben Thackray was finishing his own Ph.D. Ben's work provided an excellent starting point for my own research, and his help with measurements in the early days of the project was greatly appreciated. Vasyl Kravets has been a guide and mentor through all aspects of clean room and measurement work. I am greatly indebted to his patience and selflessness on any number of occasions throughout the Ph.D. It has been a pleasure to collaborate on so many projects with Gregory Auton, who has always been exceptionally generous with his time, helping out with e-beam lithography and 2D material work. Fran Rodriguez, Owen Marshall, Dmytro Kundys and Yang Su have all helped me out at various stages with measurements, fabrication and general pearls of wisdom.

My time in the clean room has transitioned from initial unadulterated terror to a genuine pleasure (especially on hot humid days). In addition to Vasyl, Greg and Fran, I must thank Fred Schedin, Jack Warren and Chris Berger for training, tips and general help around the clean rooms. I am also indebted to Richard O'Connor, Fran Lopez-Royo, Alex Lincoln and all others working to keep the CMN and NGI well-stocked and smoothly running.

The final year of my Ph.D. has primarily been devoted to the biosensing results described in Chap. 7. For a dyed-in-the-wool physicist, this was a strange world to enter into, but I have been very grateful to Harry Warner, Andrei Kabashin, Henri Arola, Miika Soikkeli and Philip Day for helping me feel surprisingly comfortable talking about biostuff. I must especially thank Fan Wu for her industrious help in sample fabrication and measurements.

Sasha Grigorenko has been an excellent supervisor for the past few years, providing me as necessary with corrections, guidance, corrections, suggestions, corrections and many unique conversations, while also permitting (and in some cases financing) my various travels around the world.

I must also thank Irina Grigorieva, Tom Thomson and all at the Graphene-NowNano Doctoral Training Centre for providing such a great environment in which young, foolish researchers such as myself can become slightly less young and slightly less foolish. The number of people who have provided advice, patiently endured my rants and/or organised super-amazing conferences is far too great to mention here.

I am very grateful for having survived the Ph.D. process and must thank my friends, family and the good saints of Holy Trinity Platt for help in preserving my sanity.

And finally, I thank you, dear reader, for having read this far. This thesis is as long as *A Christmas Carol*, and although this thesis is lacking in sentimental talk of goodwill to all, it does at least contain four fewer ghosts than the aforementioned novella.

Contents

Abbreviations

AOI	Angle of incidence
ATR	Attenuated total reflectance
FTIR	Fourier transform infrared spectroscopy
FWHM	Full width at half maximum
HPWG	Hybrid plasmon waveguide
IPA	Isopropyl alcohol
LSPR	Localised surface plasmon resonance
NA	Numerical aperture
NS	Nanostripe
SEM	Scanning electron microscopy
SPP	Surface plasmon polariton
SPR	Surface plasmon resonance
TE	Transverse electric
TM	Transverse magnetic
VASE	Variable angle spectroscopic ellipsometry

Chapter 1
Introduction

Metallic nanoparticles have been known for their unusual optical properties for millennia [1]. They were used extensively in Medieval times to add colour to glass in stained-glass windows [2]. This effect was even exploited as far back as in Roman times, perhaps most famously in the Lycurgus Cup which dates back to the fourth century AD [3]. Despite common awareness of this phenomenon for centuries no attempt was made to understand its origin, although Faraday made some preliminary observations of suspensions of gold particles in solution during an 1857 lecture [4].

It was not until the start of the 20th century that the first quantitative descriptions of metallic nanoparticles were published. To understand the colours of gold particles in colloidal suspensions Mie developed an electrodynamic theory of the scattering and absorption of light by spheres [5]. Although this work was largely ignored until around 1945 [6] it has come to be regarded as a seminal paper in the field of nano-optics. The theory of electromagnetic scattering by homogeneous, isotropic spheres is commonly referred to as Mie theory, although it is sometimes referred to as Lorenz-Mie-Debye theory [7] in recognition of the important contributions made by Lorenz [8][1] and Debye [9].

Meanwhile, in 1902 Wood observed anomalous absorption bands in light reflected from gratings [10]. Two types of anomaly were identified: the first were sharp anomalies caused by light beams diffracting at a grazing angle to the grating [11]; the second set were more diffuse and their origin less clear. It was understood in a theory developed by Fano in 1941 [12] (and refined by others, most notably Hessell and Oliner in 1965 [13]) that the more diffuse Wood anomalies could be described by resonant surface waves.

In 1956 Pines [14] introduced the term *plasmon* to describe collective oscillations of electrons in a study of energy losses experienced by high-energy electrons moving

[1]Lorenz published calculations of the scattering of electromagnetic radiation by spheres very similar to those of Mie in 1890. His work went largely unnoticed at the time in part because it was originally published in Danish.

P. A. Thomas, *Narrow Plasmon Resonances in Hybrid Systems*,
Springer Theses, https://doi.org/10.1007/978-3-319-97526-9_1

through metals. In this original quantum mechanical context plasmons are considered as quasiparticles arising from the quantisation of plasma oscillations. Surface waves propagating along metallic surfaces had been studied since the turn of the 20th century [15, 16]—indeed, Fano's theory describing the Wood anomalies was inspired by these works—and in 1957 Ritchie [17] was the first to consider *surface plasmons* in a theoretical study of electron energy losses on the surfaces on thin films. Surface plasmons were first experimentally observed in energy-loss experiments by Powell and Swan in 1960 [18].

The connection between surface plasmons and the resonant surface waves of Wood's anomaly was made by Ritchie et al. in 1968 [19]. In 1970 Kreibig and Zacharias showed that the optical absorption and scattering effects in metallic nanoparticles described by Mie theory could be explained as the effect of surface plasmon resonances in the nanoparticles [20].

Up until the late 1960s surface plasmons in thin metal films had been studied by bombarding the thin films with high-energy electrons. In 1968, Otto [21] and Kretchmann and Raether [22] showed that it was possible to excite surface plasmons in thin metal films using visible light via prism coupling. These prism coupling techniques made surface plasmon experiments more easily accessible to researchers and resulted in a substantial increase in surface plasmon research activity [23]. The term *surface plasmon polariton* (SPP) was introduced by Cunningham et al. [24] in 1974 to describe modes of surface plasmons; this term has since become interchangeable with "surface plasmon" in the literature [25]. The fundamental origins of SPPs in thin metal films is described in detail in Chap. 2.

The first application of surface plasmon physics came in the 1970s. In 1974 Fleischmann et al. [26] discovered that Raman scattering in pyridine molecules could be greatly enhanced when adsorbed on to the surface of a rough silver electrode. Fleischmann et al. initially attributed this enhancement to the electrode's higher surface area caused by its roughness; however, it was later recognised that this was not by itself enough to explain the magnitude of Raman signal enhancement [27, 28]. After some years of debate it was concluded that the dominant cause of this enhancement was due to the strong electric fields arising from surface plasmons [29–31]. Surface-enhanced Raman spectroscopy (SERS) has since become an important analytical technique in chemistry and surface science research [32].

The second major application of surface plasmon physics came in the 1980s. In the late 1970s it was recognised that the sensitivity of surface plasmons to their local dielectric environment could be exploited to characterise the growth of films [33] and electrochemical processes [34]. Nylander and Liedberg extended this to show that surface plasmons could be used as the basis for sensing, first for gas in 1982 [35] and then for antibodies in solution in 1983 [36]. These experiments paved the way for the first commercial surface plasmon resonance (SPR) sensors released by Biacore in 1990 [37]. Since then, SPR sensing has established itself as an important tool in the characterisation of biomolecular interactions [38], and in 2006 Biacore was acquired by General Electric for $390 million [39].

By the late 1990s advances in nanofabrication techniques, electromagnetic simulation codes and physical characterisation techniques combined to accelerate surface plasmon research output [23]. Major advances were made in both SERS [40] and SPR biosensing [41] and entirely new areas of surface plasmon research began to develop. Nanolithography allowed for the design of nanoscale waveguides which utilise SPPs to confine light to subwavelength regions [42]; it was in this context that the term *plasmonics* was first proposed [43]. It also became possible to carefully tune the geometric parameters of metallic nanoparticles and to fabricate them in regular arrays [44], allowing for new phenomena such as the diffraction coupling of localised surface plasmon resonances (LSPRs) [45, 46][2] and quantised optical transparency [47]. Plasmonic nanostructures have also been important components in the development of metamaterials: plasmonic nanoparticles can be subwavelength which allows for tuning of the overall effective permittivity and permeability of the metamaterial [48]. Such nanostructures could allow for the development of negative index materials [49] which could perhaps most notably be used in a so-called perfect lens capable of resolving subwavelength objects [50]. Modern plasmonics research has also made significant contributions in the areas of nano-optical tweezing [51], photovoltaic devices [52] and cancer therapy [53].

Most recently, there has been a trend towards studying plasmonic nanostructures in hybrid systems [42, 54], particularly with atomically thin materials such as graphene [55]. The optical properties of graphene and hexagonal boron nitride, the two dimensional materials studied in this thesis, are described in Chap. 3. Although still in a stage of relative infancy, the combination of plasmonic nanostructures and graphene has yielded significant advances in photodetection [56, 57], sensing [58, 59] and modulation [60].

It is within this context that the work of this thesis should be viewed. Spectrally narrow plasmon resonances are of particular interest because they can unlock unprecedented levels of sensitivity to their local environment. In Chaps. 4 and 7, narrow plasmon resonances are obtained using LSPRs in gold nanoparticle arrays and with SPPs in copper thin films respectively. In the remaining chapters of this thesis the extreme sensitivity of these plasmonic nanostructures is exploited by combining them with other systems to explore their potential applications. In Chap. 5 gold nanoarrays are combined with a graphene/hexagonal boron nitride heterostructure to create a nanomechanical electro-optical modulator. In Chap. 6 the same gold nanoarrays are combined with a slab of dielectric (hafnium(IV) oxide) to create a waveguide structure. In Chap. 7 copper thin films are protected with a sheet of graphene and assessed for use in biosensing.

Taken as a whole, this thesis highlights the extreme versatility of plasmonic nanostructures. Strong plasmonic effects can be attained with either nanoparticle arrays or a planar geometry. They can be combined with a variety of other materials (two-dimensional materials, waveguiding dielectrics and biological materials) to create complicated systems holding rich physics. Such hybrid systems can find applications across a wide range of disciplines, from electronics engineering to biochemical studies. The applications studied in this work emphasise the extraordinary potential of subwavelength nanostructures.

---

[2]The physical origin of diffraction coupling of LSPRs is given in Chap. 4.

References

1. P. Colomban, The use of metal nanoparticles to produce yellow, red and iridescent colour, from bronze age to present times in lustre pottery and glass: solid state chemistry, spectroscopy and nanostructure. J. Nano Res. **8**, 109–132 (2009). Trans Tech Publications, 2009
2. P. Colomban, A. Tournie, P. Ricciardi, Raman spectroscopy of copper nanoparticle-containing glass matrices: ancient red stained-glass windows. J. Raman Spectrosc. **40**(12), 1949–1955 (2009)
3. I. Freestone, N. Meeks, M. Sax, C. Higgitt, The Lycurgus cup—a Roman nanotechnology. Gold Bull. **40**(4), 270–277 (2007)
4. M. Faraday, The Bakerian lecture: experimental relations of gold (and other metals) to light. Philos. Trans. R. Soc. Lond. **147**, 145–181 (1857)
5. G. Mie, Beiträge zur Optik trüber Medien, speziell kolloidaler Metallösungen. Ann. Phys. **330**(3), 377–445 (1908)
6. M. Cardona, W. Marx, Verwechselt, vergessen, wieder gefunden. Phys. J. **11**, 27–29 (2004)
7. T. Wriedt, Mie theory: a review, in *The Mie Theory* (Springer, 2012), pp. 53–71
8. L. Lorenz, Det kongelige danske videnskabernes selskabs skrifter **6**. raekke, 6. bind, 1 (1890)
9. P. Debye, Der lichtdruck auf kugeln von beliebigem material. Ann. Phys. **335**(11), 57–136 (1909)
10. R.W. Wood, XLII. On a remarkable case of uneven distribution of light in a diffraction grating spectrum. Lond. Edinb. Dublin Philos. Mag. J. Sci. **4**(21), 396–402 (1902)
11. L. Rayleigh, On the dynamical theory of gratings. *Proc. R. Soc. Lond. Ser. A, Contain. Pap. Math. Phys. Character* **79**(532), 399–416 (1907)
12. U. Fano, The theory of anomalous diffraction gratings and of quasi-stationary waves on metallic surfaces (Sommerfelds waves). J. Opt. Soc. Am. **31**(3), 213–222 (1941)
13. A. Hessel, A.A. Oliner, A new theory of woods anomalies on optical gratings. Appl. Opt. **4**(10), 1275–1297 (1965)
14. D. Pines, Collective energy losses in solids. Rev. Mod. Phys. **28**(3), 184 (1956)
15. A. Sommerfeld, Ueber die fortpflanzung elektrodynamischer wellen längs eines drahtes. Ann. Phys. **303**(2), 233–290 (1899)
16. J. Zenneck, Über die fortpflanzung ebener elektromagnetischer wellen längs einer ebenen leiterfläche und ihre beziehung zur drahtlosen telegraphie. Ann. Phys. **328**(10), 846–866 (1907)
17. R.H. Ritchie, Plasma losses by fast electrons in thin films. Phys. Rev. **106**(5), 874 (1957)
18. C.J. Powell, J.B. Swan, Effect of oxidation on the characteristic loss spectra of aluminum and magnesium. Phys. Rev. **118**(3), 640 (1960)
19. R.H. Ritchie, E.T. Arakawa, J.J. Cowan, R.N. Hamm, Surface-plasmon resonance effect in grating diffraction. Phys. Rev. Lett. **21**(22), 1530 (1968)
20. U. Kreibig, P. Zacharias, Surface plasma resonances in small spherical silver and gold particles. Z. Phys. **231**(2), 128–143 (1970)
21. A. Otto, Excitation of nonradiative surface plasma waves in silver by the method of frustrated total reflection. Z. Phys. **216**(4), 398–410 (1968)
22. E. Kretschmann, H. Raether, Notizen: radiative decay of non radiative surface plasmons excited by light. Z. Naturforsch. A **23**(12), 2135–2136 (1968)
23. M.L. Brongersma, P.G. Kik, *Surface Plasmon Nanophotonics* (Springer, 2007)
24. S.L. Cunningham, A.A. Maradudin, R.F. Wallis, Effect of a charge layer on the surface-plasmon-polariton dispersion curve. Phys. Rev. B **10**(8), 3342 (1974)
25. H. Raether, *Surface Plasmons on Smooth Surfaces* (Springer, 1988)
26. M. Fleischmann, P.J. Hendra, A.J. McQuillan, Raman spectra of pyridine adsorbed at a silver electrode. Chem. Phys. Lett. **26**(2), 163–166 (1974)
27. D.L. Jeanmaire, R.P. Van Duyne, Surface raman spectroelectrochemistry: Part I. Heterocyclic, aromatic, and aliphatic amines adsorbed on the anodized silver electrode. J. Electroanal. Chem. Interfacial Electrochem. **84**(1), 1–20 (1977)
28. M.G. Albrecht, J.A. Creighton, Anomalously intense raman spectra of pyridine at a silver electrode. J. Am. Chem. Soc. **99**(15), 5215–5217 (1977)

29. H. Metiu, P. Das, The electromagnetic theory of surface enhanced spectroscopy. Annu. Rev. Phys. Chem. **35**(1), 507–536 (1984)
30. M. Kerker, Electromagnetic model for surface-enhanced raman scattering (SERS) on metal colloids. Acc. Chem. Res. **17**(8), 271–277 (1984)
31. M. Moskovits, Surface-enhanced spectroscopy. Rev. Mod. Phys. **57**(3), 783 (1985)
32. R.L. Garrell, Surface-enhanced Raman spectroscopy. Anal. Chem. **61**(6), 401A–411A (1989)
33. I. Pockrand, J.D. Swalen, J.G. Gordon, M.R. Philpott, Surface plasmon spectroscopy of organic monolayer assemblies. Surf. Sci. **74**(1), 237–244 (1978)
34. J.G. Gordon, S. Ernst, Surface plasmons as a probe of the electrochemical interface. Surf. Sci. **101**(1–3), 499–506 (1980)
35. C. Nylander, B. Liedberg, T. Lind, Gas detection by means of surface plasmon resonance. Sens. Actuators **3**, 79–88 (1982)
36. B. Liedberg, C. Nylander, I. Lunström, Surface plasmon resonance for gas detection and biosensing. Sens. Actuators **4**, 299–304 (1983)
37. B. Liedberg, C. Nylander, and I. Lundström, Biosensing with surface plasmon resonance–how it all started. *Biosens. Bioelectron.* **10**(8), i–ix (1995)
38. J. Homola, S.S. Yee, G. Gauglitz, Surface plasmon resonance sensors: review. Sens. Actuators B Chem. **54**(1), 3–15 (1999)
39. J. Bouley, GE Healthcare acquires Biacore Intl. for $390 million. *DDN-News* **7**(7), E070601 (2006)
40. K. Kneipp, Y. Wang, H. Kneipp, L.T. Perelman, I. Itzkan, R.R. Dasari, M.S. Feld, Single molecule detection using surface-enhanced Raman scattering (SERS). Phys. Rev. Lett. **78**(9), 1667 (1997)
41. J.N. Anker, W.P. Hall, O. Lyandres, N.C. Shah, J. Zhao, R.P. Van Duyne, Biosensing with plasmonic nanosensors. Nat. Mater. **7**(6), 442–453 (2008)
42. D.K. Gramotnev, S.I. Bozhevolnyi, Plasmonics beyond the diffraction limit. Nat. Photonics **4**(2), 83–91 (2010)
43. S.A. Maier, M.L. Brongersma, P.G. Kik, S. Meltzer, A.A.G. Requicha, H.A. Atwater, Plasmonics—a route to nanoscale optical devices. Adv. Mater. **13**(19), 1501–1505 (2001)
44. V.G. Kravets, F. Schedin, G. Pisano, B. Thackray, P.A. Thomas, A.N. Grigorenko, Nanoparticle arrays: from magnetic response to coupled plasmon resonances. Phys. Rev. B **90**(12), 125445 (2014)
45. B. Auguié, W.L. Barnes, Collective resonances in gold nanoparticle arrays. Phys. Rev. Lett. **101**(14), 143902 (2008)
46. V.G. Kravets, F. Schedin, A.N. Grigorenko, Extremely narrow plasmon resonances based on diffraction coupling of localized plasmons in arrays of metallic nanoparticles. Phys. Rev. Lett. **101**(8), 087403 (2008)
47. V.G. Kravets, F. Schedin, A.N. Grigorenko, Fine structure constant and quantized optical transparency of plasmonic nanoarrays. Nat. Commun. **3**, 640 (2012)
48. A.N. Grigorenko, A.K. Geim, H.F. Gleeson, Y. Zhang, A.A. Firsov, I.Y. Khrushchev, J. Petrovic, Nanofabricated media with negative permeability at visible frequencies. Nature **438**(7066), 335–338 (2005)
49. V.G. Veselago, The electrodynamics of substances with simultaneously negative values of ϵ and μ. Sov. Phys. Usp. **10**(4), 509 (1968)
50. J.B. Pendry, Negative refraction makes a perfect lens. Phys. Rev. Lett. **85**(18), 3966 (2000)
51. M.L. Juan, M. Righini, R. Quidant, Plasmon nano-optical tweezers. Nat. Photonics **5**(6), 349–356 (2011)
52. H.A. Atwater, A. Polman, Plasmonics for improved photovoltaic devices. Nat. Mater. **9**(3), 205–213 (2010)
53. X. Huang, I.H. El-Sayed, W. Qian, M.A. El-Sayed, Cancer cell imaging and photothermal therapy in the near-infrared region by using gold nanorods. J. Am. Chem. Soc. **128**(6), 2115–2120 (2006)
54. R.F. Oulton, V.J. Sorger, D.A. Genov, D.F.P. Pile, X. Zhang, A hybrid plasmonic waveguide for subwavelength confinement and long-range propagation. Nat. Photonics **2**(8), 496–500 (2008)

55. A.N. Grigorenko, M. Polini, K.S. Novoselov, Graphene plasmonics. Nat. Photonics **6**(11), 749–758 (2012)
56. T.J. Echtermeyer, L. Britnell, P.K. Jasnos, A. Lombardo, R.V. Gorbachev, A.N. Grigorenko, A.K. Geim, A.C. Ferrari, K.S. Novoselov, Strong plasmonic enhancement of photovoltage in graphene. Nat. Commun. **2**, 458 (2011)
57. L. Britnell, R.M. Ribeiro, A. Eckmann, R. Jalil, B.D. Belle, A. Mishchenko, Y.-J. Kim, R.V. Gorbachev, T. Georgiou, S.V. Morozov, A.N. Grigorenko, A.K. Geim, C. Casiraghi, A.N. Castro, A.H. Castro Neto, K.S. Novoselov, Strong light-matter interactions in heterostructures of atomically thin films. Science **340**(6138), 1311–1314 (2013)
58. V.G. Kravets, F. Schedin, R. Jalil, L. Britnell, R.V. Gorbachev, D. Ansell, B. Thackray, K.S. Novoselov, A.K. Geim, A.V. Kabashin, A.N. Grigorenko, Singular phase nano-optics in plasmonic metamaterials for label-free single-molecule detection. Nat. Mater. **12**(4), 304–309 (2013)
59. V.G. Kravets, R. Jalil, Y.-J. Kim, D. Ansell, D.E. Aznakayeva, B. Thackray, L. Britnell, B.D. Belle, F. Withers, I.P. Radko, Z. Han, S.I. Bozhevolnyi, K.S. Novoselov, A.K. Geim, A.N. Grigorenko, Graphene-protected copper and silver plasmonics. *Sci. Rep.* **4** (2014)
60. D. Ansell, I.P. Radko, Z. Han, F.J. Rodriguez, S.I. Bozhevolnyi, A.N. Grigorenko, Hybrid graphene plasmonic waveguide modulators. *Nat. Commun.* **6** (2015)

Chapter 2
Plasmonics

2.1 Metal Optics

We start our discussion of the fundamental origins of plasmonics by considering the behaviour of electromagnetic waves in metals. Our approach in Sects. 2.1–2.3 largely follows those of Maier [1] and Raether [2]. SI notation is used throughout.

2.1.1 Maxwell's Equations

We start by defining Maxwell's equations in matter:

$$\nabla \cdot \mathbf{D} = \rho_{ext} \tag{2.1a}$$

$$\nabla \cdot \mathbf{B} = 0 \tag{2.1b}$$

$$\nabla \times \mathbf{E} = -\frac{\partial \mathbf{B}}{\partial t} \tag{2.1c}$$

$$\nabla \times \mathbf{H} = \mathbf{J}_{ext} + \frac{\partial \mathbf{D}}{\partial t}. \tag{2.1d}$$

D is the electric displacement field, **E** is the electric field, **B** is the magnetic flux density, **H** is the magnetic field, ρ_{ext} is the external charge density and $\mathbf{J}_{ext}$ is the external current density. Note that we have followed Maier [1] in dividing the charge and current densities into their internal (ρ, $\mathbf{J}$) and external (ρ_{ext}, $\mathbf{J}_{ext}$) components such that the total charge and current densities are $\rho_{tot} = \rho_{ext} + \rho$ and $\mathbf{J}_{tot} = \mathbf{J}_{ext} + \mathbf{J}$ respectively. The external quantities drive the system and the internal quantities are responses to external stimuli.

The four microscopic fields are related to the polarisation **P** and magnetisation **M** by

$$\mathbf{D} = \epsilon_0 \mathbf{E} + \mathbf{P} \tag{2.2}$$

P. A. Thomas, *Narrow Plasmon Resonances in Hybrid Systems*, Springer Theses, https://doi.org/10.1007/978-3-319-97526-9_2

$$\mathbf{H} = \frac{1}{\mu_0}\mathbf{B} - \mathbf{M}, \tag{2.3}$$

where $\epsilon_0 = 8.854 \times 10^{-12}$ F m^{-1} (farads per metre) is the electric permittivity of free space and $\mu_0 = 4\pi \times 10^{-7}$ N A^{-2} (newtons per square ampère) is the magnetic permeability of free space.

We shall limit ourselves to linear, isotropic, nonmagnetic media. In this case we can define the following constitutive relations:

$$\mathbf{D} = \epsilon_r \epsilon_0 \mathbf{E} \tag{2.4}$$

$$\mathbf{B} = \mu_r \mu_0 \mathbf{H}. \tag{2.5}$$

ϵ_r and μ_r are respectively the relative permittivity and permeability of the medium. For convenience we will often combine the relative and free space values of permittivity and permeability as $\epsilon = \epsilon_r \epsilon_0$ and $\mu = \mu_r \mu_0$. The relative permittivity is also known as the dielectric function $\epsilon_r(\mathbf{k}, \omega)$ where $\mathbf{k}$ is the wave vector and ω is the angular frequency. In metal optics this is simplified to the limit of a spatially local response where $\mathbf{k} = 0$ and so $\epsilon_r(\mathbf{k}, \omega) = \epsilon_r(\omega)$. $\mu_r = 1$ for nonmagnetic media.

If we define the dielectric susceptibility χ such that $\epsilon_r = 1 + \chi$ we have a linear relationship between polarisation and electric field:

$$\mathbf{P} = \epsilon_0 \chi \mathbf{E}. \tag{2.6}$$

Additionally, we note that the internal current density and electric field are defined by the conductivity σ:

$$\mathbf{J} = \sigma \mathbf{E}. \tag{2.7}$$

2.1.2 Drude Model

The optical response of noble metals typically used in plasmonics is well-described by the Drude model [3, 4]. The Drude model treats metal as a series of fixed positive ions surrounded by free conduction electrons. It assumes collisions between electrons and ions or other electrons occur with a collision frequency $\gamma = 1/\tau$ (where τ is the relaxation time of a free electron gas, typically 10^{-14} s at room temperature [1]) but neglects long-range electron-electron and electron-ion interactions.

When an external electromagnetic field $\mathbf{E} = \mathbf{E}_0 e^{-i\omega t}$ is applied a free electron in the metal will move around the fixed ions with a restoring force $\mathbf{F}$ given by

$$\mathbf{F} = -e\mathbf{E} = m\ddot{\mathbf{x}} + m\gamma\dot{\mathbf{x}} + m\omega_0^2\mathbf{x}, \tag{2.8}$$

where $\mathbf{x}$ is the displacement of electrons from the ions, $e = 1.602 \times 10^{-19}$ C is the elementary charge, m is the effective electron mass and ω_0 is the frequency of the restoring force. In the case of the Drude model we assume there is no restoring force acting on the electrons so we set $\omega_0 = 0$. Trying the solution $\mathbf{x} = \mathbf{x}_0 e^{-i\omega t}$ yields

$$\mathbf{x} = \frac{e}{m}\frac{1}{\omega^2 + i\omega\gamma}\mathbf{E}. \tag{2.9}$$

This is a complex amplitude: the imaginary component describes any phase shifts between the driving electric field and the response of the electrons. The polarisation of the metal caused by the displacement of electrons is given by $\mathbf{P} = -ne\mathbf{x}$ (where n is the number of electrons), which when combined with Eqs. 2.6, 2.9 and the definition $\epsilon_r = 1 + \chi$ gives the dielectric function of the metal:

$$\epsilon_r(\omega) = 1 - \frac{\omega_p^2}{\omega^2 + i\omega\gamma}. \tag{2.10}$$

Here we have introduced the plasma frequency of the metal $\omega_p^2 = \frac{ne^2}{\epsilon_0 m}$. For $\omega < \omega_p$ metals retain their character and there is negligible damping, which allows us to simplify the dielectric function to

$$\epsilon_r(\omega) = 1 - \frac{\omega_p^2}{\omega^2}. \tag{2.11}$$

For $\omega > \omega_p$ we must make a correction for noble metals. In this regime the free-electron model gives $\epsilon \to 1$, but in noble metals the free s electrons are the dominant factor in the response. The d band is close to the Fermi surface, leading to a high level of polarisation due to the positive ion cores. To account for this we must add an extra dielectric constant ϵ_∞ (where typically $1 \leq \epsilon_\infty \leq 10$) to Eq. 2.11 to give

$$\epsilon_r(\omega) = \epsilon_\infty - \frac{\omega_p^2}{\omega^2 + i\omega\gamma}. \tag{2.12}$$

Although the Drude model is entirely classical, and for example makes no consideration of Fermi-Dirac statistics, it remains a surprisingly accurate model [5]. The main limitation of the Drude model is that it does not describe the interband transitions that occur in the visible spectrum for many metals such as gold. In this thesis, however, we will be primarily concerned with the optical response of metals in the near-mid infrared spectral region, where the Drude model does indeed provide a very accurate model of the optical response of noble metals.

2.1.3 Electromagnetic Waves

Maxwell's equations can easily be solved to give the three-dimensional wave equations for electromagnetic waves propagating through a dielectric. In this case there are no external charges or currents and we can combine (2.1c) and (2.1d) to give

$$\nabla \times (\nabla \times \mathbf{E}) = -\mu \frac{\partial^2 \mathbf{D}}{\partial t^2} \tag{2.13}$$

$$\nabla (\nabla \cdot \mathbf{E}) - \nabla^2 \mathbf{E} = -\mu \frac{\partial^2 \mathbf{D}}{\partial t^2}. \tag{2.14}$$

Here we have used the vector identity $\nabla \times (\nabla \times \mathbf{A}) = \nabla (\nabla \cdot \mathbf{A}) - \nabla^2 \mathbf{A}$. By noting from Eq. 2.1a that $\nabla \cdot \mathbf{E} = 0$

$$\nabla^2 \mathbf{E} = \epsilon\mu \frac{\partial^2 \mathbf{E}}{\partial t^2} \tag{2.15a}$$

$$\nabla^2 \mathbf{B} = \epsilon\mu \frac{\partial^2 \mathbf{B}}{\partial t^2}. \tag{2.15b}$$

This allows us to identify the speed of light in a dielectric as $v = 1/\sqrt{\epsilon\mu}$. The solutions for these equations are

$$\mathbf{E} = \mathbf{E}_0 e^{i(\mathbf{k}\cdot\mathbf{r}-\omega t)} \tag{2.16}$$

$$\mathbf{B} = \mathbf{B}_0 e^{i(\mathbf{k}\cdot\mathbf{r}-\omega t)}. \tag{2.17}$$

Substituting this back into Eq. 2.14 and assuming a nonmagnetic medium gives

$$\mathbf{k}(\mathbf{k} \cdot \mathbf{E}) - k^2 \mathbf{E} = -\epsilon_r(\omega) \frac{\omega^2}{c^2}, \tag{2.18}$$

where $c = 1/\sqrt{\epsilon_0\mu_0}$ is the speed of light in a vacuum. For transverse waves $\mathbf{k} \cdot \mathbf{E} = 0$ and the dispersion relation is

$$k^2 = \epsilon_r(\omega) \frac{\omega^2}{c^2}. \tag{2.19}$$

For longitudinal waves the left hand side of Eq. 2.18 becomes zero, requiring $\epsilon_r(\omega) = 0$. Longitudinal modes correspond to volume plasmons (which fall outside the scope of this thesis), so therefore only occur for frequencies at which the dielectric function is zero.

We also note that substituting Eq. 2.16 into Eq. 2.14 gives the Helmholtz equation:

$$\nabla^2 \mathbf{E} + k_0^2 \epsilon_r \mathbf{E} = 0, \tag{2.20}$$

where $k_0 = \frac{\omega}{c}$.

It is at the interface between a dielectric and a metal that things become more complicated and surface plasmons start to appear.

2.2 Surface Plasmon Polaritons

2.2.1 Origin

Let us consider a one-dimensional metal-dielectric interface as illustrated in Fig. 2.1. Waves will propagate at the interface in the x direction with no variation in the y direction. The plane of the propagating wave and the metal-dielectric interface itself will occur at $z = 0$. For $z < 0$ we will have a metal with dielectric function ϵ_1, electric field $\mathbf{E}_1$ and magnetic field $\mathbf{H}_1$. For $z > 0$ we will have a dielectric with dielectric function ϵ_2, electric field $\mathbf{E}_2$ and magnetic field $\mathbf{H}_2$. We shall consider only p-polarised light since an analysis of the boundary conditions shows no surface modes exist for s-polarised light [1]. The electromagnetic fields therefore take the following form:

$$\mathbf{E}_1 = \begin{pmatrix} E_{x1} \\ 0 \\ E_{z1} \end{pmatrix} e^{i(k_{x1}x - k_{z1}z - \omega t)} \tag{2.21}$$

$$\mathbf{H}_1 = \begin{pmatrix} 0 \\ H_{y1} \\ 0 \end{pmatrix} e^{i(k_{x1}x - k_{z1}z - \omega t)} \tag{2.22}$$

$$\mathbf{E}_2 = \begin{pmatrix} E_{x2} \\ 0 \\ E_{z2} \end{pmatrix} e^{i(k_{x2}x + k_{z2}z - \omega t)} \tag{2.23}$$

$$\mathbf{H}_2 = \begin{pmatrix} 0 \\ H_{y2} \\ 0 \end{pmatrix} e^{i(k_{x2}x + k_{z2}z - \omega t)}. \tag{2.24}$$

These fields must satisfy the following boundary conditions:

$$E_{x1} = E_{x2} = E_x \tag{2.25}$$

$$H_{y2} = H_{y1} = H_y \tag{2.26}$$

$$\epsilon_1 E_{z1} = \epsilon_2 E_{z2}. \tag{2.27}$$

The first two boundary conditions lead to $k_{x1} = k_{x2} = k_x$. Assuming no external currents we can use Maxwell's fourth equation (Eq. 2.1d) to find the dispersion relation for surface plasmons:

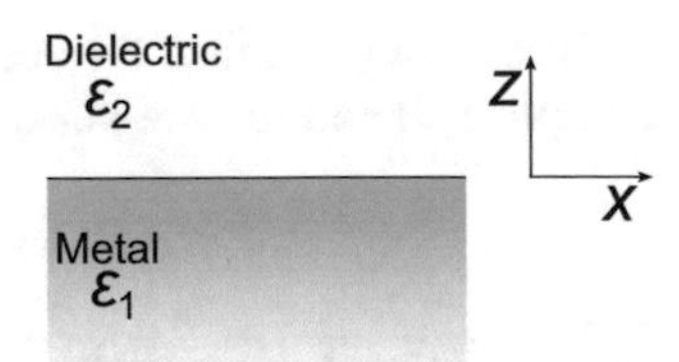

Fig. 2.1 Definition of metal-dielectric boundary used to derive SPP dispersion relation

$$\frac{k_{z1}}{\epsilon_1} + \frac{k_{z2}}{\epsilon_2} = 0. \tag{2.28}$$

Since $k_x^2 + k_y^2 + k_z^2 = k^2$ and in our case $k_y = 0$ the Drude dispersion relation becomes

$$k_x^2 + k_{zi}^2 = \epsilon_i \frac{\omega^2}{c^2}. \tag{2.29}$$

Substituting Eq. 2.28 into this yields

$$k_x = \frac{\omega}{c}\sqrt{\frac{\epsilon_1 \epsilon_2}{\epsilon_1 + \epsilon_2}}. \tag{2.30}$$

For the case of surface plasmons excited at an air-metal interface there will be $\epsilon_2 = 1$ while $\epsilon_1 < 0$. This means k_x will be complex and so therefore lossy, with exponential decay in the z-direction.

We note that in practice surface plasmons are most commonly excited in metallic thin films in an asymmetric environment (such as glass/metal/air). In these cases it is possible to derive the dispersion relation for surface plasmons using Fresnel's equations for reflection and transmission at multiple boundaries.

2.2.2 *Excitation Methods*

Although the first investigations of plasmons relied on excitation using high-energy electrons [6], studies of SPPs now almost universally rely on SPP excitation using light. Figure 2.2 shows the dispersion relations of SPPs and free space photons (with $\omega = ck_x$). The lack of any intersection between the two dispersion relations (except for $\omega = 0$) means that free space photons cannot directly excite SPPs ($k_x > \omega/c$). We must instead use a coupling method.

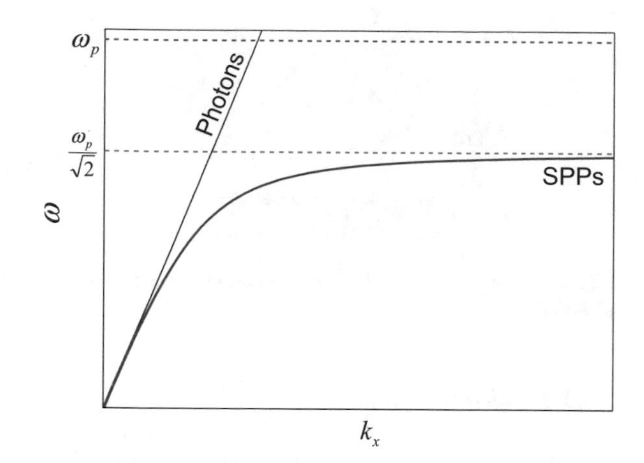

Fig. 2.2 The dispersion relations for SPPs and photons. The lack of intersection between these lines makes direct excitation of SPPs by free space photons impossible

2.2.2.1 Prism Coupling

One coupling method, known at the attenuated total reflectance (ATR) method, involves changing the momentum of the incident light by passing it through a dielectric prism (usually glass or quartz). The momentum projection of the light on to the surface of the metal becomes

$$k_x = \sqrt{\epsilon_0}\frac{\omega}{c}\sin\theta_0, \tag{2.31}$$

where ϵ_0 is the dielectric constant of the prism and θ_0 is the incident angle of the light beam inside the prism.[1] Increasing the dielectric constant changes the gradient of the photon dispersion line in Fig. 2.2 so that it overlaps with the SPP dispersion curve. Increasing ϵ_0 will also change the gradient of the SPP dispersion curve by the same factor, so prism coupling can only excite surface plasmons on a metal/air interface.

Two ATR configurations exist: the Otto configuration [7] (Fig. 2.3a) consists of a prism/air/metal film system. It is used in studies where the surface of interest may be damaged by direct contact with the prism. The Kretschmann configuration [8] (Fig. 2.3b) is a prism/metal/air system. It much more commonly used than the Otto configuration due to its relative simplicity. A metal film can either be deposited directly on to the bottom of the prism or on to a glass wafer. If it is deposited on a separate glass wafer an index-matching fluid such as glycerol must be added between the prism and the bare glass side of the wafer.

[1]Note that here ϵ_0 refers to the dielectric constant for the prism, not the electric permittivity of free space, to provide labelling continuity with the metal (ϵ_1) and dielectric (ϵ_2).

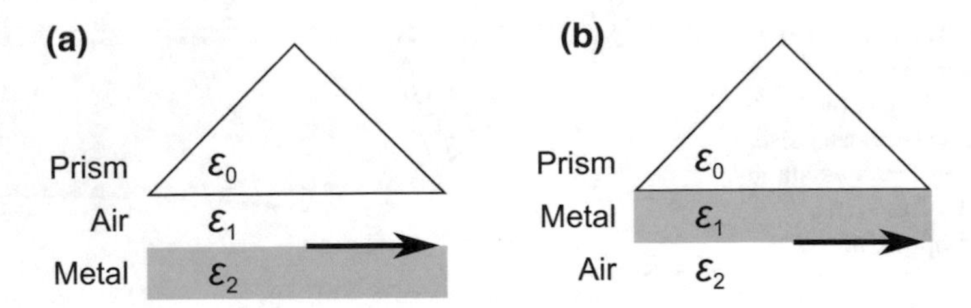

Fig. 2.3 The **a** Otto and **b** Kretschmann configurations for prism coupling of light to surface plasmons

2.2.2.2 Grating Coupling

It is also possible to excite surface plasmons on metal surfaces using a grating. Light incident on a surface at an angle of incidence of θ_0 will have a component in the surface with wavevector $k_x = \frac{\omega}{c}\sin\theta_0$. If the surface has a grating with period a then the wavevector for the component of light on the surface will be

$$k_x = \frac{\omega}{c}\sin\theta_0 \pm \Delta k_x, \tag{2.32}$$

where

$$\Delta k_x = \frac{2\pi\nu}{a} \tag{2.33}$$

and ν is an integer. This acts to shift the photon dispersion curve in Fig. 2.2 so that it intersects the SPP dispersion curve. Grating coupling can act to both couple and decouple light from a metal surface and is a common feature in many waveguide designs [1, 9].

The effect of prism and grating coupling on the photon dispersion curve is shown in Fig. 2.4.

2.2.3 Controlling the Properties of Surface Plasmon Polaritons

For certain applications it is important to ensure that the surface plasmon resonance occurs at a specific wavelength. For example, waveguides and modulators designed for telecommunication applications must utilise plasmon resonances that occur at a wavelength of 1.5 μm. Other applications such as sensing are best served by plasmon resonances that are particularly narrow. It is possible to tune these properties by varying certain design parameters.

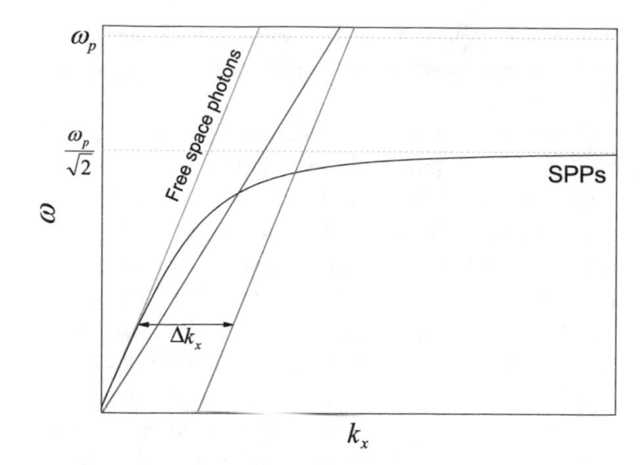

Fig. 2.4 The effect of prism (blue line) and grating (green line) coupling on the free space photon dispersion curve (grey line)

2.2.3.1 Metallic Film Thickness

Fresnel's equations for a three-layer system provide a quantitative description for the reflection of incident light in plasmonic ATR systems. The reflection R for p-polarised light in an asymmetric environment is

$$R = |r_{012}^p|^2 = \left|\frac{E_r^p}{E_0^p}\right|^2 = \left|\frac{r_{01}^p + r_{12}^2 e^{2ik_{z1}d}}{1 + r_{01}^p r_{12}^p e^{2ik_{z1}d}}\right|^2, \tag{2.34}$$

where E_0^p and E_r^p are the electric fields of the incident and reflected fields respectively, d is the thickness of the metal film and

$$r_{ik}^p = \left(\frac{k_{zi}}{\epsilon_i} - \frac{k_{zk}}{\epsilon_k}\right) \cdot \left(\frac{k_{zl}}{\epsilon_i} + \frac{k_{zk}}{\epsilon_k}\right)^{-1}. \tag{2.35}$$

When $|Re(\epsilon_1)| \gg 1$ and $|Im(\epsilon_1)| \ll |Re(\epsilon_1)|$ this can be approximated to give a Lorentzian-type equation:

$$R = 1 - \frac{4\Gamma_i \Gamma_{rad}}{[k_x - (k_x^0 + \Delta k_x)]^2 + (\Gamma_i + \Gamma_{rad})^2}. \tag{2.36}$$

This gives the resonance wavevector as $k_x^0 + \Delta k_x$ which differs from Eq. 2.30 since we are now considering metal film of finite thickness instead of the semi-infinite thickness of the metal film treated in Eq. 2.30. For $e^{2ik_x d} \ll 1$, Δk_x can approximately be given as

$$\Delta k_x = \left[\frac{\omega}{c}\frac{2}{1 + |Re(\epsilon_1)|}\left(\frac{|Re(\epsilon_1)|}{|Re(\epsilon_1)| - 1}\right)^{\frac{3}{2}} e^{-2|k_x^0|d}\right] r_{01}^p(k_x^0). \tag{2.37}$$

The real component of Δk_x shifts the resonance position compared to 2.30 and the imaginary part gives a damping term Γ_{rad} in addition to the internal damping $\Gamma_i = \text{Im}(k_x^0)$.

The reflection reaches a minimum ($R = 0$) when $\Gamma_i = \Gamma_{rad}$. Physically, we consider that light passes through initial glass dielectric and is partially reflected at glass/metal boundary but mostly transmitted into the metal layer. Once travelling inside the metal layer it will decay exponentially until it reaches the metal/air boundary where it will excite surface plasmons. Some light at the metal/air interface will be reradiated back through the metal from the metal/air boundary excitations. This backscattered field is in anti-phase with the incident field, so these two fields can potentially cancel each other out. Increasing the film thickness causes the backscattered field to decrease in strength and decreasing the film thickness causes it to increase. Choosing the correct film thickness therefore allows one to minimise the reflection, potentially so that the incident and backscattered fields completely cancel each other out, yielding $R = 0$.

2.2.3.2 Angle of Incidence

The simplest way of fine-tuning the spectral position of a surface plasmon resonance is by changing the angle of incidence of the light beam. Figure 2.4 shows that there is usually only one point at which the photon dispersion and SPP dispersion curves intersect (sometimes two in the case of prism coupling). Therefore, if $k_x = n\frac{\omega}{c}\sin\theta_0$ (where n is the ambient refractive index) and θ_0 is changed, the SPP resonance must then occur at a different frequency. This effect is relatively small for surface plasmon resonances on thin films (see Fig. 2.5b). However, in nanoparticle arrays the optical response can dramatically change with angle of incidence as different plasmon resonance modes within each nanoparticle are excited in different planes. This effect is used most dramatically in Chap. 4.

2.2.3.3 Dielectric Environment

Replacing the air at the metal-air interface with a different dielectric will alter the SPP dispersion curve in Fig. 2.4. A material with higher dielectric constant will act to decrease the gradient of the SPP dispersion curve, decreasing the frequency of incident light (and so increasing the wavelength) needed to excite SPPs. This sensitivity to changes at the metal-dielectric interface is what has made surface plasmons such an interesting tool for sensing applications.

2.2.3.4 Choice of Metal

The strongest influence on the plasmonic response of a thin film is the type of metal used. In general, metals with high electron densities and few or no interband

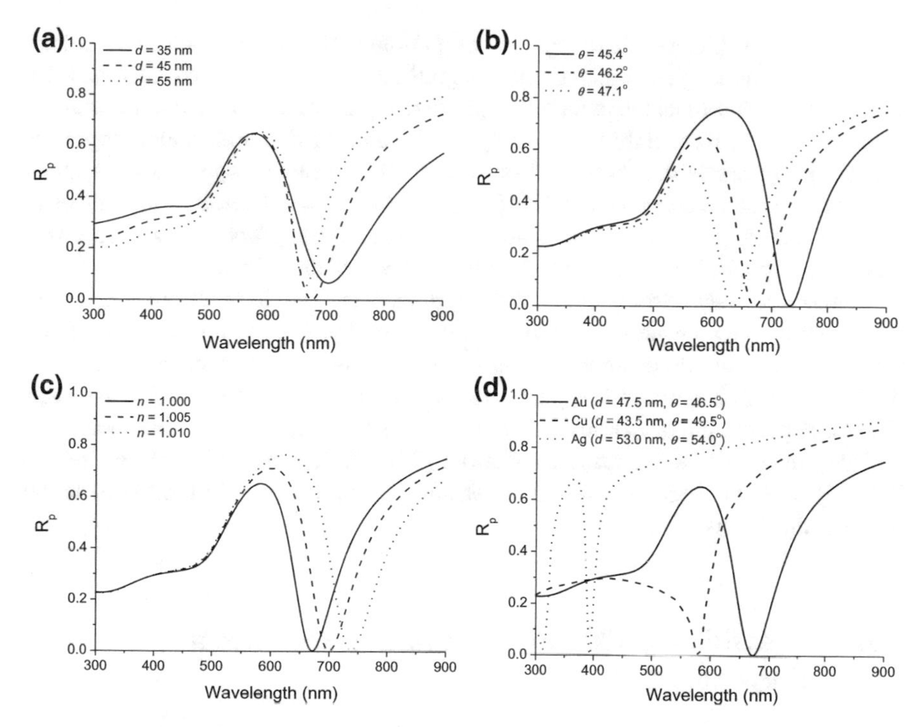

Fig. 2.5 Fresnel modelling of surface plasmon resonances in the Kretschmann configuration for a film of gold, thickness 47.5 nm, angle of incidence 46.2° and $n = 1$ with variations in **a** film thickness d, **b** angle of incidence θ, **c** ambient refractive index n and **d** choice of metal

transitions are the strongest candidates for plasmonic metals. According to these criteria silver is the best candidate for plasmonics since it has the highest conductivity of any metal and very weak interband transitions in the visible spectrum. The lack of any strong interband transitions means it is also has the least damping of any plasmonic metal. However, its reactivity in ambient conditions causes it to oxidise relatively quickly [10], limiting long-term device applications.

Gold does not have quite as high a conductivity as silver; however, it is exceptionally stable. Indeed, our plasmonic nanostructures fabricated from gold show little to no deterioration in their optical response after some years. One limitation of gold is the strong interband transitions that occur in the blue and green spectral regions (which give rise to its orange-yellow colour). Outside this region, however, gold is a strong candidate for plasmonics and is capable of producing high-quality plasmon resonances in the red [11] and near-infrared [12] regions.

Silver and gold are by far the most commonly used metals in plasmonics research [10], although copper has been highlighted as a potential competitor to these metals. A huge challenge with using gold and silver in plasmonic nanocircuitry is their incompatibility with conventional silicon manufacturing techniques: gold and silver both diffuse into silicon and can create traps which adversely affect electronic performance [13, 14]. Meanwhile, copper is compatible with semiconductor technology [15]. Copper has a higher conductivity than gold and similar interband transitions [16, 17], meaning that it has the potential to produce higher-quality plasmon resonances than gold in approximately the same spectral region [18]. Furthermore, copper is considerably cheaper than both gold and silver [10]. However, copper oxidises much more rapidly than either gold or silver [19], leading to rapid deterioration of its plasmon resonances [20]. A potential solution to this is to use graphene as a protective barrier on copper [21], an idea which will explore further in Chap. 7.

Some metals have strong interband transitions in the visible spectrum but possess strong plasmon resonances at ultraviolet wavelengths [22]. Aluminium has received a considerable amount of interest in recent years [23–25], but other metals such as rhodium [26, 27], gallium [28] and indium [29] have also been studied for their ultraviolet plasmonic properties. In all these cases oxidation is a limiting factor.

Other materials such as semiconductors [30], alloys [31] and superconductors [32, 33] have also been considered for plasmonic applications, but fall outside the scope of this thesis.

2.2.4 *Deposition of Thin Films Using Electron-Beam Evaporation*

All metal films studied in this thesis were deposited using electron-beam evaporation with glass substrates of thickness 1 mm. Briefly, cleaned substrates are placed in an evacuated chamber along with metal pellets. The metal pellets are placed in a crucible and heated with an electron beam. The metal then evaporates and is deposited on surfaces; the deposition rate on the substrates is primarily controlled by the distance between the crucible and the substrates and the current and position of the electron beam on the metal pellets. A metallic sublayer of chromium is always deposited before depositing the main film of interest to promote adhesion between the glass substrate and the metal thin film.

2.2.5 *Characterisation of Thin Films Using Spectroscopic Ellipsometry*

Our primary tool for characterisation of thin films (and, indeed, all our plasmonic nanostructures) was spectroscopic ellipsometry [34]. Ellipsometry measures the complex reflectance ratio of p- and s-polarised light in terms of the parameters Ψ and Δ, connected by the expression

$$\frac{r_p}{r_s} = \tan(\Psi)e^{i\Delta}, \tag{2.38}$$

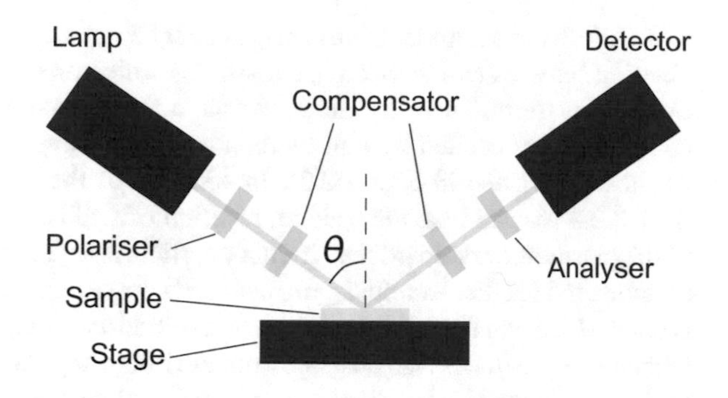

Fig. 2.6 Schematic of variable angle spectroscopic ellipsometer in reflection mode

where r_p and r_s are Fresnel's reflection coefficients for p- and s-polarised light respectively. Ψ describes the amplitude of reflected light and Δ the phase difference. Ellipsometry measures a ratio which eliminates a lot of noise, making it a highly sensitive measuring technique.

Ellipsometry is most commonly used to determine the optical (n, k and ϵ_r) and physical (thickness, roughness) properties of thin layered films. This is achieved by taking a series of spectra at different angles of incidence and fitting the spectra to an appropriate model (e.g. Fresnel coefficients, effective medium theories or Mueller-Jones matrices).

Figure 2.6 shows the general design schematic of an ellipsometer. White light (in our case from a xenon lamp) is irradiated through a polariser (to polarise the light) and then a compensator (to modify the polarisation state of the light) before being reflected from (or transmitted through) the sample. The reflected light passes through a compensator and analyser (to determine the polarisation state of reflected light) before reaching the detector.

We used a Woollam M-2000F ellipsometer which collected optical spectra from 245 to 1688 nm at incident angles in the range 45–90°. In addition to measuring Ψ and Δ our ellipsometer was also used to measure raw reflection and transmission data for p- and s-polarised light (R_p, R_s, T_p and T_s respectively).

The unfocussed spot was approximately 3–5 mm in diameter depending on the angle of incidence. When studying smaller regions it is possible to use focussing optics with numerical aperture 0.1, reducing the spot diameter to approximately 50 μm. The use of focussing optics does introduce a degree of spatial decoherence [35], which can act to limit certain long-range effects in plasmonic nanoarrays.

2.3 Localised Surface Plasmon Resonances

We have so far only considered surface plasmons in planar metal films; however, modern plasmonics research is predominantly focused on the study of localised surface plasmon resonances (LSPRs) in metallic nanostructures [36–39].

LSPRs have a number of advantages over SPPs in continuous metal films. LSPRs occur in subwavelength metal nanoparticles which means they are no longer translationally invariant. Therefore, no k_x conservation is required which means that LSPRs can be directly excited by light without the use of any coupling mechanisms (such as those described in Sect. 2.2.2). In addition to the tuning methods described in Sect. 2.2.3 we can tune the spectral position of LSPRs by changing the dimensions of the plasmonic nanoparticles. If a nanoparticle is asymmetric it can contain multiple resonant modes. For example, gold nanorods possess two distinct plasmon resonance modes which can be excited by electron oscillations along the major or minor axis of the nanorod [40]. It is also possible to create long-range coupling effects in plasmonic nanoarrays by tuning the geometry of nanoparticle arrays [11].

2.3.1 The Frölich Condition and Mie Theory

We can describe the interaction of electromagnetic radiation with nanoparticles using the quasistatic approximation. This approach is valid if we assume the nanoparticles are perfect spheres with radius a that can be described by a dielectric function ϵ_1 surrounded by a dielectric with dielectric constant ϵ_2. We assume a quasistatic constant external electric field $\mathbf{E} = E_0\hat{\mathbf{z}}$ and that the diameter of the nanoparticles is sufficiently subwavelength that the phase of the harmonically oscillating electromagnetic field is approximately constant over the volume of the nanoparticle.

To describe the optical response we must determine the potential Φ from which we can find electric field ($\mathbf{E} = -\nabla\Phi$). To do this we solve the Laplace equation $\nabla^2\Phi = 0$. The solution takes the form of a summation of Legendre polynomials, which after applying boundary conditions gives the internal and external potentials (Φ_{in} and Φ_{out} respectively) as

$$\Phi_{\text{in}} = -\frac{3\epsilon_2}{\epsilon_1+2\epsilon_2} E_0 r \cos\theta \tag{2.39a}$$

$$\Phi_{\text{out}} = -E_0 r \cos\theta + \frac{\epsilon_1-\epsilon_2}{\epsilon_1+2\epsilon_2} E_0 a^3 \frac{\cos\theta}{r^2}. \tag{2.39b}$$

Equation 2.39b represents the external potential due to the external applied field (the first term) and the dipole located at the centre of the particle (the second term). This allows us to introduce the dipole moment $\mathbf{p}$ into the second term in Eq. 2.39b:

$$\Phi_{\text{out}} = -E_0 r \cos\theta + \frac{\mathbf{p}\cdot\mathbf{r}}{4\pi\epsilon_0\epsilon_2 r^3} \tag{2.40}$$

$$\mathbf{p} = 4\pi\epsilon_0\epsilon_2 a^3 \frac{\epsilon_1 - \epsilon_2}{\epsilon_1 + 2\epsilon_2}\mathbf{E}_0. \tag{2.41}$$

If we define the polarisability α such that $\mathbf{p} = \epsilon_0\epsilon_2\alpha\mathbf{E}_0$ we see that the polarisability of a subwavelength sphere in the electrostatic approximation is

$$\alpha = 4\pi a^3 \frac{\epsilon_1 - \epsilon_2}{\epsilon_1 + 2\epsilon_2}. \tag{2.42}$$

When $\mathrm{Im}[\epsilon_1]$ is small or varies slowly the maximum polarisability (and so maximum field enhancement) occurs when

$$\mathrm{Re}\,[\epsilon_1] = -2\epsilon_2. \tag{2.43}$$

This is known as the Frölich condition and is associated with the dipole surface plasmon of a metallic nanoparticle in an oscillating field. For a Drude metal the resonance occurs when $\omega_0 = \omega_p/\sqrt{3}$. ω_0 is highly dependent on ϵ_2 and will redshift with increasing ϵ_2. The magnitude of field enhancement and the width of resonance are limited by damping (represented by the imaginary part of ϵ_1).

The above quasistatic approach is only valid for vanishingly small particles, although provides a reasonable approximation for spherical and ellipsoidal particles with diameters less than 100 nm. For nanoparticles any larger than this, however, there will still be significant changes in the phase of the electric field over the volume of the particle. In this case an electrodynamic approach is necessary which can be provided by Mie theory. Mie theory expands the internal and scattered fields in terms of a set of normal modes described by vector harmonics. The Frölich condition can be recovered from the Mie absorption and scattering coefficients by taking their power series expansions and retaining only the first term.

2.3.2 *Fabrication of Plasmonic Nanoarrays Using Electron-Beam Lithography*

Plasmonic nanoparticles can be created using either bottom-up or top-down fabrication methods. Bottom-up approaches rely on chemical synthesis methods and are in general more efficient and cheaper than top-down methods. It is possible to create ordered nanoparticle arrays either using colloidal self-assembly (if the nanoparticles are assembled on a specially-prepared surface as in Ref. [41]) or nanosphere lithography [42], although in general the precision of array geometries compared to top-down approaches is limited. Self-assembly techniques have, however, recently been used to create plasmonic nanoarrays with nontrivial optical responses including topological darkness [43] and optical magnetism [44], highlighting the increasing sophistication of self-assembly techniques.

In contrast to bottom-up fabrication, top-down fabrication is a subtractive processes, meaning that material is removed to create the required structure. This is more expensive and less efficient than bottom-up fabrication but allows for much more precise control of the size, shape and location of features. For us it was crucial to ensure that our nanoparticle arrays has the correct dimensions and spacing

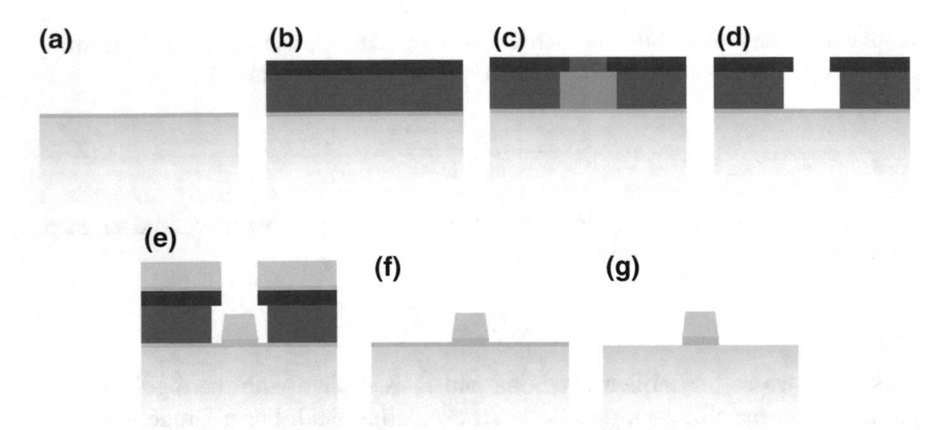

Fig. 2.7 Overview of electron-beam lithography using glass substrates. (Dimensions not to scale.) **a** Deposition of Cr anti-charging sublayer. **b** Spin-coating of bilayer resist. **c** Exposure of resist to scanning electron beam. **d** Development of exposed pattern. **e** Deposition of Cr adhesion sublayer and thin film of primary metal (in this thesis gold). **f** Lift-off. **g** Cr etch

in order to create long-range collective effects. We therefore used top-down fabrication techniques; because our nanostructures were subwavelength in size we used electron-beam lithography.

Briefly, in electron-beam lithography a polymer resist is spin-coated on to a substrate and then a pattern is created in the resist by exposing parts of it to an electron beam. Exposing the resist to the electron beam chemically alters the resist so that the solubility of the exposed resist is either higher (in the case of positive resists) or lower (in the case of negative resists) than the unexposed regions. After exposure the sample is placed in a developing solution which either dissolves the exposed parts (for positive resist) or the unexposed parts (for negative resists). A metal thin film is then evaporated on the sample (see Sect. 2.2.4). The metal is either deposited directly on the substrate (where the resist has been developed) or on top of the undeveloped resist. The thickness of resist is chosen such that the metal on the substrate and metal on the resist are unconnected. Finally, the remaining resist is then dissolved, leaving the desired metallic nanostructure adhered to the substrate. Because the undesired metal is "lifted off" the substrate by the dissolving of the resist this particular step is known as "lift-off".

The technical details for our e-beam lithography procedure are as follows (illustrated in Fig. 2.7):

1. Prepare substrate. We used $25 \times 25\ \text{mm}^2$ glass substrates of thickness 1 mm cut from $75 \times 25\ \text{mm}^2$ ultraclean microscope slides. These were first cleaned by sonication in acetone and isopropyl alcohol (IPA) for 15 min each. Glass substrates are insulating which will cause electrostatic build-up on the surface. To prevent charging by the electron beam a thin chromium sublayer (approximately 3 nm) was deposited on the substrate by e-beam evaporation.

2. Prepare resist. We use a positive bilayer resist where the top layer will slightly overhang the bottom layer after exposure, preventing unwanted connection between the metal deposited on the substrate and on top of the resist. The first layer is poly(methyl methacrylate) (PMMA) molecular weight 495 K in a 3% solution of anisole (by mass) spin-coated at a speed of 3000 rpm for 60 s. The second layer is PMMA 950K 2% spin-coated at a speed of 5000 rpm for 60 s. Each layer is baked for 5 min on a hot plate at approximately 150 °C.
3. Write desired pattern into resist using scanning electron microscope beam.
4. Develop resist by placing in a solution of MIBK:IPA 1:3 for 30 s then rinsing in IPA for a further 30 s.
5. Desired metal thin film is deposited by electron-beam evaporation (Sect. 2.2.4). A chromium sublayer is deposited to improve adhesion between the metal film and substrate.
6. Lift-off: the sample is left in heated acetone until all the PMMA dissolves, unwanted metal is lifted off and only the desired nanostructure is left adhered to the substrate.
7. The presence of the chromium sublayer deposited in step 1 can potentially hinder measurements (for example, by shorting electrical measurements or suppressing plasmon resonances by resistive coupling [45]). If this was likely to be an issue for measurements the sample was placed in a chromium etch solution for 3–5 s to remove the chromium sublayer.

2.4 Factors Affecting the Shape of Plasmon Resonances

In Sect. 2.2.3 we discussed a variety of experimental parameters that can be tuned to adjust the spectral position of the plasmon resonance. Here we discuss the factors affecting the lineshape of plasmon resonances.

2.4.1 Losses

In Sect. 2.1.2 we introduced the Drude model for the dielectric function of a metal (Eq. 2.12). This is a complex function that can be separated into real and imaginary components:

$$\mathrm{Re}\left[\epsilon_r(\omega)\right] = \epsilon_\infty - \frac{\omega_p^2}{\omega^2 + \gamma^2} \tag{2.44}$$

$$\mathrm{Im}\left[\epsilon_r(\omega)\right] = \frac{\omega_p^2 \gamma}{\omega^3 + \omega\gamma^2}. \tag{2.45}$$

ϵ_∞, ω_p and γ are all intrinsic properties of the metal. Plasmon resonances can occur in the negative band of Re[$\epsilon_r(\omega)$]; the frequency range of this band is determined by all three parameters. Im[$\epsilon_r(\omega)$] describes the losses in the metal, dependent on ω, ω_p and γ. The idea of optical losses increasing with plasma frequency $\omega_p^2 = \frac{ne^2}{\epsilon_0 m}$ is an intuitive one, since higher ω_p arises from a greater density of free electrons n capable of interacting with electromagnetic radiation while potentially also possessing lower effective mass m.

The collision frequency $\gamma = 1/\tau$ describes the time it takes for the electrons to respond to the incident electromagnetic field. It is common to assume that τ is a constant independent of the frequency of incident light as is the case for the classical Drude model, which assumes that between electron-electron and electron-ion collisions the trajectories of electrons are straight [5]. This means that electron-electron and electron-ion interactions between collisions are neglected (the independent electron and free electron approximations, respectively).

In reality, however, $\tau = \tau(\omega)$ is a function of ω [46–49]:

$$\frac{1}{\tau(\omega)} = \frac{1}{\tau_0} + b\omega^2, \tag{2.46}$$

where τ_0 is the optical relaxation time at zero frequency and b is a constant.

A number of factors have been suggested that contribute towards the frequency dependence of $\tau(\omega)$. The frequency dependence of $\tau(\omega)$ suggests that free electrons in noble metals behave more like a Fermi liquid than a Fermi gas [49]. Gurzhi [50] showed that electron-electron interactions in a Fermi liquid of electrons give rise to a contribution τ_{ee} to $\tau(\omega)$ that is also dependent on the temperature of the Fermi liquid:

$$\frac{1}{\tau_{ee}(\omega)} = \frac{1}{\tau_{ee}^0}\left[1 + \left(\frac{\hbar\omega^2}{2\pi k_B T}\right)^2\right], \tag{2.47}$$

where k_B is Boltzmann's constant and T is the temperature. However, the frequency dependence of $\tau(\omega)$ predicted by electron-electron scattering is too small when compared to experimental data to be considered a leading contributor to this dependence [47, 48].

The number of impurities and grain boundaries in a metal film has a substantial effect on $\tau(\omega)$. Indeed, annealing a metal thin film has been shown to substantially increase its $\tau(\omega)$ [46, 47]. Electrons in a metal film will move to screen impurities or grain boundaries, meaning that a film with larger grains and fewer impurities will have a higher $\tau(\omega)$.

Electron-phonon scattering gives the largest contribution to b. Normal electron-phonon scattering modes (where momentum is conserved) contribute a very small fourth-order frequency dependence but provide no second-order frequency dependence of $\tau(\omega)$ [48]. However, umklapp scattering provides a significant contribution. In umklapp scattering an electron or phonon is scattered such that the particle's wavevector is transformed outside the Brillouin zone (i.e. momentum is not conserved) [5]. The rate of umklapp scattering is determined by the Fermi level of the electrons in the metal (and therefore the temperature of the metal) and also on the energy of incident radiation [48]. $\tau(\omega)$ thus decreases with increasing frequency.

The quality factor Q of a Lorentzian resonance centred at frequency ω_0 is defined by

$$Q = \frac{\omega_0}{\gamma} = \omega_0 \tau(\omega). \tag{2.48}$$

We therefore expect higher-quality plasmon resonances to occur in metals with fewer scattering events i.e. there are fewer impurities, grain boundaries and umklapp events.

Of the three noble metals Au has the highest $\tau(\omega) = (1.6 + 0.09\omega^2) \times 10^{14}\ \mathrm{s}^{-1}$ while Ag has the smallest $\tau(\omega) = (1.2 - 0.09\omega^2) \times 10^{14}\ \mathrm{s}^{-1}$. Copper has $\tau(\omega) = (1.43 + 0.15\omega^2) \times 10^{14}\ \mathrm{s}^{-1}$ [48]. Therefore, the plasmon resonances in silver are higher quality than those of copper, which in turn are higher than those of gold.

2.4.2 *Fano Asymmetry*

When a sharp, discrete resonance and a broad spectral line or continuum occur in the same spectral region they will interfere constructively and destructively at different spectral points. This results in an asymmetric resonance known as a Fano resonance [51] of the form

$$I \alpha \frac{(F\gamma + \omega - \omega_0)^2}{(\omega - \omega_0)^2 + \gamma^2}, \tag{2.49}$$

where I is the intensity of the measured response and F is the Fano parameter, describing the degree of asymmetry in the resonance.

Since Fano resonances rely on interference between two optical features they are prominently observed in nanoparticle arrays. For example, the plasmon resonances described in Chaps. 4–6 arise when sharp, narrow Wood anomalies occur in the same spectral region are broader localised plasmon resonances, resulting in Fano-like resonances. Any Fano-like asymmetry in surface plasmon polaritons on thin metal films is the result of interplay between the plasmon resonance and interband transitions as in Fig. 7.2.

References

1. S.A. Maier, *Plasmonics: Fundamentals and Applications* (Springer Science & Business Media, 2007)
2. H. Raether, *Surface Plasmons on Smooth Surfaces* (Springer, 1988)
3. P. Drude, Zur elektronentheorie der metalle. Ann. Phys. **306**(3), 566–613 (1900)

4. P. Drude, Zur elektronentheorie der metalle; II. Teil. galvanomagnetische und thermomagnetische effecte. Ann. Phys. **308**(11), 369–402 (1900)
5. N.W. Ashcroft, N.D. Mermin, *Solid State Physics* (Harcourt Brace, Orlando, 1976)
6. C.J. Powell, J.B. Swan, Effect of oxidation on the characteristic loss spectra of aluminum and magnesium. Phys. Rev. **118**(3), 640 (1960)
7. A. Otto, Excitation of nonradiative surface plasma waves in silver by the method of frustrated total reflection. Z. Phys. **216**(4), 398–410 (1968)
8. E. Kretschmann, H. Raether, Notizen: radiative decay of non radiative surface plasmons excited by light. Z. Naturforsch. A **23**(12), 2135–2136 (1968)
9. E. Devaux, T.W. Ebbesen, J.-C. Weeber, A. Dereux, Launching and decoupling surface plasmons via micro-gratings. Appl. Phys. Lett. **83**(24), 4936–4938 (2003)
10. P.R. West, S. Ishii, G.V. Naik, N.K. Emani, V.M. Shalaev, A. Boltasseva, Searching for better plasmonic materials. Laser Photonics Rev. **4**(6), 795–808 (2010)
11. V.G. Kravets, F. Schedin, A.N. Grigorenko, Extremely narrow plasmon resonances based on diffraction coupling of localized plasmons in arrays of metallic nanoparticles. Phys. Rev. Lett. **101**(8), 087403 (2008)
12. B.D. Thackray, P.A. Thomas, G.H. Auton, F.J. Rodriguez, O.P. Marshall, V.G. Kravets, A.N. Grigorenko, Super-narrow, extremely high quality collective plasmon resonances at telecom wavelengths and their application in a hybrid graphene-plasmonic modulator. Nano Lett. **15**(5), 3519–3523 (2015)
13. G. Bemski, Recombination properties of gold in silicon. Phys. Rev. **111**(6), 1515 (1958)
14. L.D. Yau, C.T. Sah, Measurement of trapped-minority-carrier thermal emission rates from Au, Ag, and Co traps in silicon. Appl. Phys. Lett. **21**(4), 157–158 (1972)
15. G.V. Naik, V.M. Shalaev, A. Boltasseva, Alternative plasmonic materials: beyond gold and silver. Adv. Mater. **25**(24), 3264–3294 (2013)
16. B.R. Cooper, H. Ehrenreich, H.R. Philipp. Optical properties of noble metals. II. *Phys. Rev.* **138**(2A), A494 (1965)
17. P.B. Johnson, R.W. Christy, Optical constants of the noble metals. Phys. Rev. B **6**(12), 4370 (1972)
18. M. Futamata, Application of attenuated total reflection surface-plasmon-polariton raman spectroscopy to gold and copper. Appl. Opt. **36**(1), 364–375 (1997)
19. N. Tajima, M. Fukui, Y. Shintani, O. Tada, In situ studies on oxidation of copper films by using atr technique. J. Phys. Soc. Jpn. **54**(11), 4236–4240 (1985)
20. G.H. Chan, J. Zhao, E.M. Hicks, G.C. Schatz, R.P. Van Duyne, Plasmonic properties of copper nanoparticles fabricated by nanosphere lithography. Nano Lett. **7**(7), 1947–1952 (2007)
21. V.G. Kravets, R. Jalil, Y.-J. Kim, D. Ansell, D.E. Aznakayeva, B. Thackray, L. Britnell, B.D. Belle, F. Withers, I.P. Radko, Z. Han, S.I. Bozhevolnyi, K.S. Novoselov, A.K. Geim, A.N. Grigorenko, Graphene-protected copper and silver plasmonics. *Sci. Rep.* **4** (2014)
22. J.M. McMahon, G.C. Schatz, S.K. Gray, Plasmonics in the ultraviolet with the poor metals Al, Ga, In, Sn, Tl, Pb, and Bi. Phys. Chem. Chem. Phys. **15**(15), 5415–5423 (2013)
23. C. Langhammer, M. Schwind, B. Kasemo, I. Zoric, Localized surface plasmon resonances in aluminum nanodisks. Nano Lett. **8**(5), 1461–1471 (2008)
24. M.W. Knight, L. Liu, Y. Wang, L. Brown, S. Mukherjee, N.S. King, H.O. Everitt, P. Nordlander, N.J. Halas, Aluminum plasmonic nanoantennas. Nano Lett. **12**(11), 6000–6004 (2012)
25. M.W. Knight, N.S. King, L. Liu, H.O. Everitt, P. Nordlander, N.J. Halas, Aluminum for plasmonics. ACS Nano **8**(1), 834–840 (2014)
26. B. Ren, X.-F. Lin, Z.-L. Yang, G.-K. Liu, R.F. Aroca, B.-W. Mao, Z.-Q. Tian, Surface-enhanced raman scattering in the ultraviolet spectral region: UV-SERS on rhodium and ruthenium electrodes. J. Am. Chem. Soc. **125**(32), 9598–9599 (2003)
27. A.M. Watson, X. Zhang, R. Alcaraz de La Osa, J.M. Sanz, F. González, F. Moreno, G. Finkelstein, J. Liu, H.O. Everitt, Rhodium nanoparticles for ultraviolet plasmonics. Nano Lett. **15**(2), 1095–1100 (2015)
28. M.W. Knight, T. Coenen, Y. Yang, B.J.M. Brenny, M. Losurdo, A.S. Brown, H.O. Everitt, A. Polman, Gallium plasmonics: deep subwavelength spectroscopic imaging of single and interacting gallium nanoparticles. ACS Nano **9**(2), 2049–2060 (2015)

29. M.B. Ross, G.C. Schatz, Aluminum and indium plasmonic nanoantennas in the ultraviolet. J. Phys. Chem. C **118**(23), 12506–12514 (2014)
30. A. Boltasseva, H.A. Atwater, Low-loss plasmonic metamaterials. Science **331**(6015), 290–291 (2011)
31. M.G. Blaber, M.D. Arnold, M.J. Ford, A review of the optical properties of alloys and intermetallics for plasmonics. J. Phys. Condens. Matter **22**(14), 143201 (2010)
32. A. Tsiatmas, A.R. Buckingham, V.A. Fedotov, S. Wang, Y. Chen, P.A.J. De Groot, N.I. Zheludev, Superconducting plasmonics and extraordinary transmission. Appl. Phys. Lett. **97**(11), 111106 (2010)
33. V.A. Fedotov, A. Tsiatmas, J.H. Shi, R. Buckingham, P. De Groot, Y. Chen, S. Wang, N.I. Zheludev, Temperature control of fano resonances and transmission in superconducting metamaterials. Opt. Exp. **18**(9), 9015–9019 (2010)
34. H. Tompkins, E.A. Irene, *Handbook of Ellipsometry* (William Andrew, Springer, 2005)
35. B.D. Thackray, V.G. Kravets, F. Schedin, G. Auton, P.A. Thomas, A.N. Grigorenko, Narrow collective plasmon resonances in nanostructure arrays observed at normal light incidence for simplified sensing in asymmetric air and water environments. ACS Photonics **1**(11), 1116–1126 (2014)
36. H.A. Atwater, A. Polman, Plasmonics for improved photovoltaic devices. Nat. Mater. **9**(3), 205–213 (2010)
37. W.L. Barnes, A. Dereux, T.W. Ebbesen, Surface plasmon subwavelength optics. Nature **424**(6950), 824–830 (2003)
38. E. Ozbay, Plasmonics: merging photonics and electronics at nanoscale dimensions. Science **311**(5758), 189–193 (2006)
39. K.A. Willets, R.P. Van Duyne, Localized surface plasmon resonance spectroscopy and sensing. Annu. Rev. Phys. Chem. **58**, 267–297 (2007)
40. S. Link, M.A. El-Sayed, Spectral properties and relaxation dynamics of surface plasmon electronic oscillations in gold and silver nanodots and nanorods (1999)
41. P. Evans, W.R. Hendren, R. Atkinson, G.A. Wurtz, W. Dickson, A.V. Zayats, R.J. Pollard, Growth and properties of gold and nickel nanorods in thin film alumina. Nanotechnology **17**(23), 5746 (2006)
42. C.L. Haynes, R.P. Van Duyne, Nanosphere lithography: a versatile nanofabrication tool for studies of size-dependent nanoparticle optics. J. Phys. Chem. B **105**(24), 5599–5611 (2001)
43. L. Malassis, P. Massé, M. Tréguer-Delapierre, S. Mornet, P. Weisbecker, P. Barois, C.R. Simovski, V.G. Kravets, A.N. Grigorenko, Topological darkness in self-assembled plasmonic metamaterials. Adv. Mater. **26**(2), 324–330 (2014)
44. S. Gomez-Graña, A. Le Beulze, M. Treguer-Delapierre, S. Mornet, E. Duguet, E. Grana, E. Cloutet, G. Hadziioannou, J. Leng, J.-B. Salmon, V.G. Kravets, A.N. Grigorenko, N.A. Peyyety, V. Ponsinet, P. Richetti, A. Baron, D. Torrent, P. Barois, Hierarchical self-assembly of a bulk metamaterial enables isotropic magnetic permeability at optical frequencies. Mater. Horiz. **3**(6), 596–601 (2016)
45. V.G. Kravets, F. Schedin, A.N. Grigorenko, Fine structure constant and quantized optical transparency of plasmonic nanoarrays. Nat. Commun. **3**, 640 (2012)
46. M.-L. Thèye, Investigation of the optical properties of au by means of thin semitransparent films. Phys. Rev. B **2**(8), 3060 (1970)
47. S.R. Nagel, S.E. Schnatterly, Frequency dependence of the Drude relaxation time in metal films. Phys. Rev. B **9**(4), 1299 (1974)
48. J.B. Smith, H. Ehrenreich, Frequency dependence of the optical relaxation time in metals. Phys. Rev. B **25**(2), 923 (1982)
49. S.J. Youn, T.H. Rho, B.I. Min, K.S. Kim, Extended Drude model analysis of noble metals. Phys. Status Solidi (b) **244**(4), 1354–1362 (2007)
50. R.N. Gurzhi, Mutual electron correlations in metal optics. Sov. Phys. JETP **8**(4), 673–675 (1959)
51. B. Luk'yanchuk, N.I. Zheludev, S.A. Maier, N.J. Halas, P. Nordlander, H. Giessen, C.T. Chong, The Fano resonance in plasmonic nanostructures and metamaterials. Nat. Mater. **9**(9), 707–715 (2010)

Chapter 3
Two-Dimensional Materials

3.1 Introduction

The first study of an atomically thin material was published by Wallace [1] in 1947. Wallace was studying the electronic properties of graphite: he started his study by considering the electronic properties of a single layer of graphite (later termed *graphene*) and then extended this to bulk graphite. At the time graphite was regarded as an important material because of its use as a mediator in nuclear fission reactors; Wallace was not interested in atomically thin structures per se. Indeed, in the previous decade studies by Peierls [2] and Landau [3] suggested that free-standing atomically thin materials would not be thermodynamically stable. They argued that as materials become thinner atomic displacements caused by thermal fluctuations would become comparable in size to the thickness of the material which would forbid any long-range order. These ideas were backed up by a series of experiments and went unchallenged for some decades [4].

This commonly accepted notion was upended by Novoselov et al. [5] who showed that it is possible to isolate single layers of graphene from bulk graphite using micromechanical cleavage techniques. Single-layered graphene was subsequently found to possess a number of remarkable properties in addition to its stability: its charge carriers behave like massless Dirac fermions [6], its optical absorption depends only on the fine structure constant [7], its charge carrier mobility can be as high as 1,000,000 cm^2/V s [8], it possesses a tunable band gap [9]; it also possesses extraordinary mechanical [10] and thermal properties [11] and has an exceptionally high surface area [12]. These superlatives have led to the investigation of graphene in an enormous range of potential applications, including electronics [5] and optoelectronics [13], composite materials [14], energy storage [15] and biomedicine [16].

Single layered graphene is atomically thin and its electrons are confined to move in two dimensions. Therefore, graphene is known as a two-dimensional (2D) material [6] and an allotrope of carbon; in the same way, we can think of buckminsterfullerene, carbon nanotubes and bulk graphite as zero-, one- and three-dimensional allotropes of carbon, respectively.

P. A. Thomas, *Narrow Plasmon Resonances in Hybrid Systems*,
Springer Theses, https://doi.org/10.1007/978-3-319-97526-9_3

Graphene was the first 2D material to be isolated and studied, but by no means the last [6]. A huge library of dozens of 2D materials now exists with a diverse set of properties [17]. Combining these different 2D materials into Van der Waals heterostructures allows for the creation of incredibly compact unique designer materials for the study of new physics and new potential applications [18]. The combination of hexagonal boron nitride and graphene has proven especially fruitful [19], with studies of boron nitride-encapsulated graphene devices revealing a variety of novel phenomena over a decade after graphene's initial isolation [20–22].

3.2 Graphene

We will now describe the key electronic and optoelectronic properties of graphene. We will largely follow the approach of Katsnelson [23] and Neto et al. [24].

3.2.1 *Electronic Properties*

3.2.1.1 Crystal Structure

Graphene is a honeycomb crystal lattice of carbon atoms (Fig. 3.1a). The lattice is defined in real space by two lattice vectors $\mathbf{a}_1$ and $\mathbf{a}_2$:

$$\mathbf{a}_1 = \frac{a}{2}(3, \sqrt{3}), \tag{3.1}$$

$$\mathbf{a}_2 = \frac{a}{2}(3, -\sqrt{3}). \tag{3.2}$$

$a \approx 1.42$ Å is the nearest-neighbour distance. There are two atoms in each unit cell, which takes the form of a rhombus. We can also consider the honeycomb lattice as consisting of two sublattices A and B. One atom from sublattice A is surrounded by three atoms from sublattice B (and vice versa) with the following three nearest neighbour vectors:

$$\boldsymbol{\delta}_1 = \frac{a}{2}(1, \sqrt{3}), \tag{3.3}$$

$$\boldsymbol{\delta}_2 = \frac{a}{2}(1, -\sqrt{3}), \tag{3.4}$$

$$\boldsymbol{\delta}_3 = \frac{a}{2}(-1, 0). \tag{3.5}$$

The reciprocal lattice of graphene is shown in Fig. 3.1b, which also shows the Brillouin zone. The reciprocal lattice vectors are

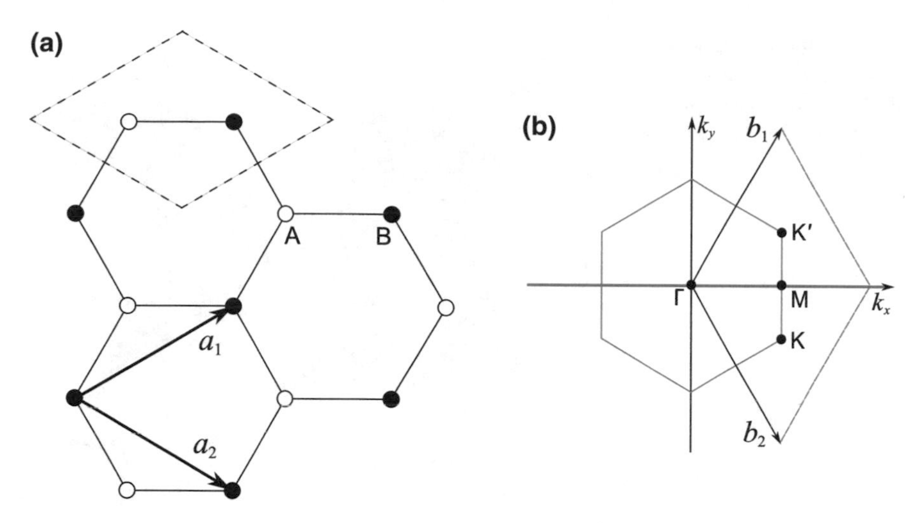

Fig. 3.1 Graphene honeycomb lattice. **a** Real space lattice showing lattice vectors $\mathbf{a}_1$ and $\mathbf{a}_2$, the A and B sublattices and the unit cell (dashed rhombus). **b** Reciprocal lattice vectors $\mathbf{b}_1$ and $\mathbf{b}_2$, the edge of the first Brillouin zone (solid hexagon) and some important points

$$\mathbf{b}_1 = \frac{2\pi}{3a}\left(1, \sqrt{3}\right), \tag{3.6}$$

$$\mathbf{b}_2 = \frac{2\pi}{3a}\left(1, -\sqrt{3}\right). \tag{3.7}$$

The carbon atoms are covalently bonded: three of carbon's four valence electrons from each carbon atom form three σ bonds. These in-plane bonds are the strongest form of covalent bond and response for graphene's extraordinary strength. The nearest-neighbour distance $a \approx 1.42$ Å is between the length of a single C C bond ($r \approx 1.54$ Å) and double C=C bond ($r \approx 1.31$ Å) and sufficiently small that the out-of-plane p-orbitals overlap, leading to the formation of π bonds. The two p-orbitals that hybridise to form a π bond can either be in phase (forming lower-energy π bonds) or out of phase (forming higher-energy π^* bonds). Each π bond can hold two electrons (one spin up, one spin down), but since there is one electron in each constituent p-orbital the lower energy π bond is filled and the higher-energy π^* bond is empty. When all π bonds in the graphene are considered together as bands we see that the π band acts as the filled valence band and the π^* band as the conduction band.

3.2.1.2 Band Structure

The band structure of graphene has been calculated using the tight-binding model. The tight-binding model is an established model in solid state physics used to calculate the band structure of materials [25]. It assumes that the atomic lattice can be

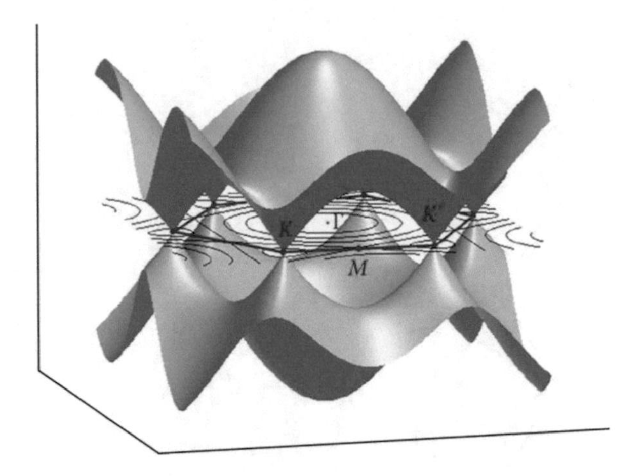

Fig. 3.2 Band structure of graphene calculated using the tight-binding model. Reproduced from [26]

described by a series of atomic wavefunctions with small corrections to account for the small overlap of adjacent atomic wavefunctions. We assume the bound levels of the Hamiltonian are well-localised so that the wavefunction of one atom $\psi(\mathbf{r})$ becomes small when $|\mathbf{r}| \to a$. Then we can add hopping parameters to account for the probability of electrons in one atom moving to the nearest neighbour t (and even next-nearest neighbour, t').

Taking into account the nearest and next-nearest neighbours, the tight-binding model gives the band structure of graphene as

$$E(\mathbf{k}) = \pm\sqrt{3 + f(\mathbf{k})} + t' f(\mathbf{k}), \tag{3.8}$$

where

$$f(\mathbf{k}) = 2\cos\left(\sqrt{3}k_y a\right) + 4\cos\left(\frac{\sqrt{3}}{2}k_y a\right)\cos\left(\frac{3}{2}k_x a\right), \tag{3.9}$$

$t = 2.7$ eV and $t' = -0.2t$. If we only considered the nearest-neighbour hopping parameter the band structure would be symmetric around zero energy. The finite value of t' breaks the electron-hole symmetry so that the π and $\pi *$ bands become antisymmetric. The band structure for the first Brillouin zone is plotted in Fig. 3.2.

The valence and conduction (π and $\pi *$) bands touch at Dirac points at the K and K' points in the Brillouin zone. If we expand the full band structure close to the $\mathbf{K}$ or $\mathbf{K}'$ vector by defining $\mathbf{k} = \mathbf{K} + \mathbf{q}$ where $|\mathbf{q}| \ll |\mathbf{K}|$ one obtains

$$E_{\pm}(\mathbf{q}) \approx \pm v_F|\mathbf{q}| + O\left[(q/K)^2\right]. \tag{3.10}$$

$v_F = 3ta/2 \approx 1 \times 10^6$ m s^{-1} is the Fermi velocity. The most dramatic difference between this and the standard result of $v = \sqrt{2E/m}$ (where m is the electron's mass) is that our result is independent of mass, allowing for much greater changes of velocity with energy.

In its undoped state at $T = 0$ K the valence band is completely full and the conduction band empty: it is a semimetal with zero band gap. One can easily shift the Fermi energy E_F from the Dirac point by electric field doping [5]. Although graphene's extraordinary electronic properties made it an attractive candidate for application in logic circuits, it has not been possible to open up a band gap in graphene and retain its electronic properties while achieving a practical on-off ratio [27].

3.2.2 Optical Properties

Graphene's band structure gives rise to an unusual set of optical properties. The real part of graphene's optical conductance due to interband transitions is [28]

$$G_1(\omega) = G_0 \equiv \frac{e^2}{4\hbar} \approx 6.08 \times 10^{-5}\Omega^{-1}, \tag{3.11}$$

where $\hbar$ is Planck's constant divided by 2π. This is unusual in that it depends solely on fundamental constants and has no relation to any physical material properties. One can find the transmission of graphene by applying Fresnel's equations in the thin-film limit to give [29]

$$T = \left(1 + \frac{2\pi}{c}G\right) \tag{3.12}$$

$$= \left(1 + \frac{\pi\alpha}{2}\right)^{-2} \approx 1 - \pi\alpha \approx 0.977 \tag{3.13}$$

for light at normal incidence, where $\alpha = e^2/\hbar c \approx 1/137$ is the fine structure constant. The opacity of few-layer graphene is approximately proportional to the number of layers involved for up to around four layers [7]. Graphene was the first material observed with optical absorption determined solely by the fine structure constant, although it was later shown that this property occurs for any low-energy 2D electronic system [30], including certain plasmonic nanoarrays [31].

It was initially assumed that Eq. 3.13 would only hold for lower energies where the electronic spectrum of graphene is linear. At higher energies we expect some deviation from $\pi\alpha$ absorption due to triangular warping and nonlinearities in the band structure of graphene. However, these deviations turned out to be small (less than 2% for photons with energy up to 3 eV i.e. green light). The deviations only become significant for blue light, meaning that graphene's optical transmission remains at 2.3% from the visible through to the infrared part of the spectrum [7].

Fig. 3.3 reveals an additional absorption peak in the ultraviolet. Saddle points in the band structure of 2D crystals give rise to singularities in the density of states [32]. These are known as van Hove singularities and manifest themselves as absorption

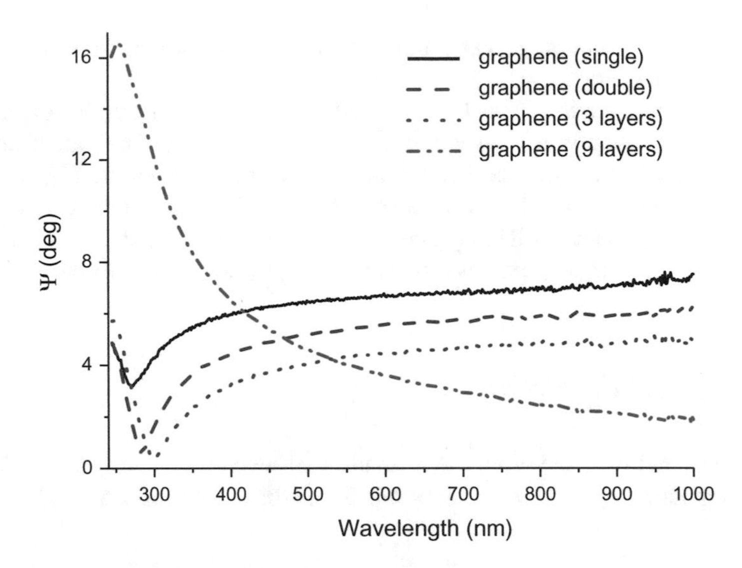

Fig. 3.3 Van Hove singularity in ellipsometric spectrum of single-layer, bilayer and few-layer graphene. Angle of incidence = 60°. The position of the van Hove peak shifts with increasing layers of graphene as the band structure changes. Spectra measured by Vasyl G. Kravets

peaks in transmission and reflection measurements and also allow for enhancement effects in Raman spectroscopy studies [33]. In graphene the M saddle point gives rise to a van Hove singularity at 4.6 eV (approximately 270 nm) [30]. This is shifted from the initially-predicted value of around 5.2 eV due to strong resonant excitonic effects which act to redistribute the optical transition strengths at these energies [34].

Shifting E_F in graphene by electrostatic gating affects not just its electrical properties but its optical properties. If one shifts E_F the states in the conduction band up to the energy E_F are occupied. The symmetry of the valence and conduction bands means that direct excitations of valence band electrons with energy $-E_F < E < 0$ to conduction band states with energy $0 < E < +E_F$ are forbidden by the Pauli exclusion principle (see Fig. 3.4). Therefore, interband optical transitions in gated graphene are forbidden with photons of energy $\hbar\omega < 2E_F$ [9, 36]. This phenomenon is known as optical Pauli blocking and provides a simple, fast way of controlling the transparency of graphene (although intraband transitions can still occur). Perhaps most significantly, it is the underlying mechanism for much of the current research in graphene optical modulator designs [37].

3.2.3 Intrinsic Graphene Plasmons

Although outside the scope of this thesis, we note the existence and interest in intrinsic graphene plasmons [38]. Graphene plasmons rely on intraband transitions close to E_F in gated graphene; the dispersion relationship of graphene plasmons can therefore be

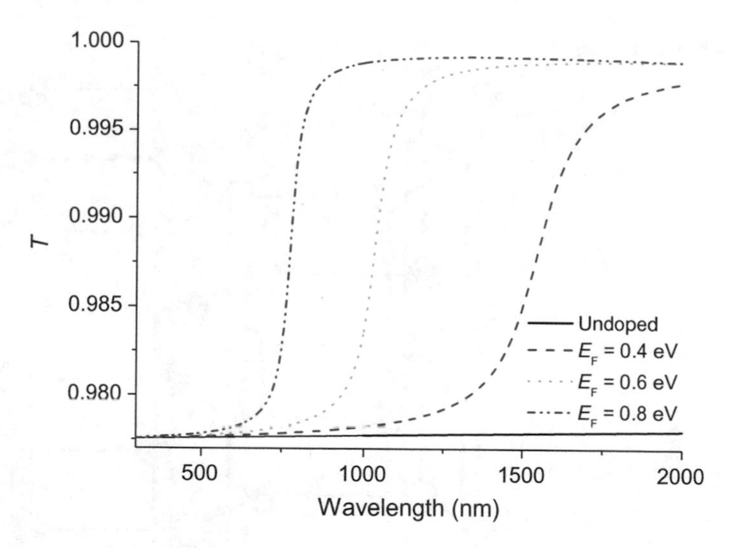

Fig. 3.4 Gate-tunable Pauli blocking of optical interband transitions in graphene: normal transmission through single-layer graphene calculated as a function of Fermi energy E_F. Calculated using the derivation as reported in [35]

tuned by shifting E_F [39]. In addition to their tunability graphene plasmons are noted for their lifetime [40], strong field confinement and low losses [41]. Applications of graphene plasmons exist largely in the terahertz and mid-infrared spectrum and include areas such as sensing, spectroscopy and security [42].

3.3 Hexagonal Boron Nitride

Hexagonal boron nitride (hBN) is one of three allotropes of boron nitride (the others being cubic [43] and wurtzite [44] boron nitride). It has a layered hexagonal structure (shown in Fig. 3.5) similar to graphite with alternating boron and nitrogen atoms. Bulk hBN is layered such that boron nitride atoms sit on top of nitrogen atoms and vice versa (AA') [45]. The B-N bond length of 1.446 Å is very close to that of graphite, giving a lattice mismatch of just 1.7% [46]. The small lattice mismatch and the ability to isolate atomically smooth hBN planes has led to hBN being used as an atomically smooth substrate for graphene [19].

hBN is an insulator with a direct band gap of over 5 eV [47, 48]. Band gap measurements in the literature have varied from 3.6 to 7.1 eV [47] due to limitations in the quality of samples. The optical response of hBN in the ultraviolet spectrum is complex due its band structure [49], but in the visible spectrum it has a transparency of over 99% [48].

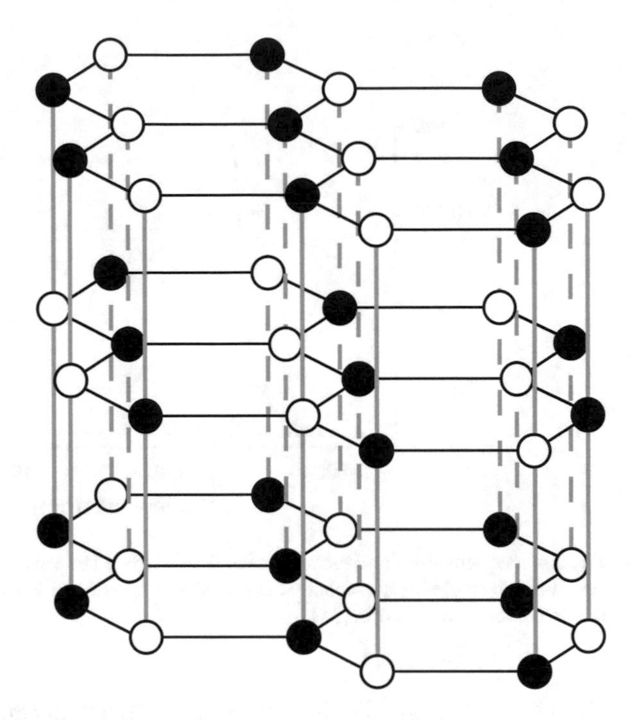

Fig. 3.5 Crystal structure of hBN

3.3.1 Reststrahlen Band in hBN

When considering the response of an ionic crystal to an external electric field we must consider both the atomic polarisability (arising from distortion of ionic charge distribution) and displacement polarisability (due to ionic displacements) of the ionic crystal. It can be shown that these conditions give rise to the following dielectric function for ionic crystals [25]:

$$\epsilon_r(\omega) = \epsilon_\infty + \frac{\epsilon_\infty - \epsilon_0}{(\omega^2/\omega_{\mathrm{TO}}^2) - 1}. \tag{3.14}$$

ϵ_0 is the static dielectric constant of the crystal, describing the behaviour of the crystal in a static electric field. ϵ_∞ is the dielectric constant of the ion at optical frequencies, related to the refractive index by $n = \sqrt{\epsilon_\infty}$. ω_{TO} is given by

$$\omega_{\mathrm{TO}}^2 = \bar{\omega}^2 \left(\frac{\epsilon_\infty + 2}{\epsilon_0 + 2} \right) = \bar{\omega}^2 \left(1 - \frac{\epsilon_0 - \epsilon_\infty}{\epsilon_0 + 2} \right), \tag{3.15}$$

where $\bar{\omega}$ is characteristic of the lattice vibrational frequencies, mathematically defined from the calculation of the displacement polarisability of the primitive cell of the lattice as $\bar{\omega}^2 = k/M$, where k is the "spring constant" of the ionic bond in the primitive cell of the lattice and M is the ionic reduced mass of the positive (M_+)

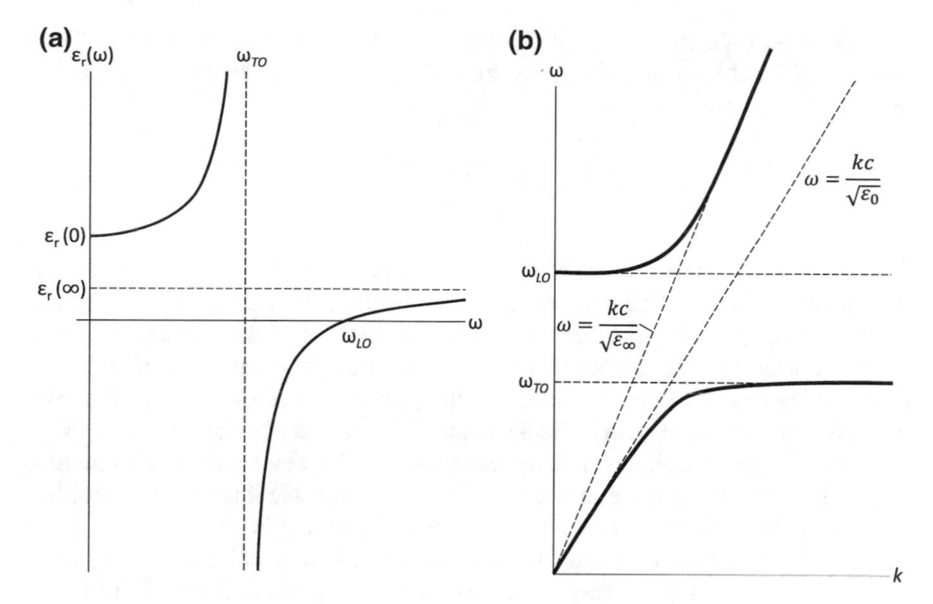

Fig. 3.6 Optical phonon modes in an insulator such as hBN. **a** Dielectric function $\epsilon_r(\omega)$ and **b** the transverse electromagnetic dispersion relation $\omega = ck/\sqrt{\epsilon_r(\omega)}$ of modes for a generic diatomic ionic crystal. Adapted from [25]

and negative (M_-) ions, $M^{-1} = (M_+)^{-1} + (M_-)^{-1}$. Equation 3.14 is comparable to Eq. 2.12 (the Drude model) for metals: while Eq. 2.12 shows the optical response of metals is fundamentally defined by the plasma frequency of free electrons in the metal, Eq. 3.14 shows that the optical response of ionic crystals is predominantly defined by the resonant frequency of lattice vibrations in the crystal known as optical phonons.

Equation 3.14 diverges when $\omega = \omega_{\mathrm{TO}}$ and corresponds to the case of the electric polarisation **P** being perpendicular to the electromagnetic wavevector **k**; ω_{TO} is hence known as the transverse optical phonon mode. When **P** and **k** are parallel $\epsilon_r = 0$; this corresponds to the longitudinal optical phonon mode ω_{LO}:

$$\omega_{\mathrm{LO}}^2 = \frac{\epsilon_0}{\epsilon_\infty}\omega_{\mathrm{TO}}^2. \tag{3.16}$$

Figure 3.6a plots the generic dielectric function of a diatomic ionic crystal (Eq. 3.14). For $\omega_{\mathrm{TO}} < \omega < \omega_{\mathrm{LO}}$ the dielectric function is negative. Transverse fields must satisfy the dispersion relation ($k^2 = \epsilon_r(\omega)\omega^2/c^2$, Eq. 2.19), meaning that ck must be imaginary for this frequency range. Therefore, no radiation in this band can propagate through the crystal.

Figure 3.6b plots the generic dispersion relation for transverse electromagnetic modes propagating in a diatomic ionic crystal such as hBN. There are two separate branches, each constrained by two asymptotes. The upper branch is constrained by the lines $\omega = \omega_{\mathrm{LO}}$ and $\omega = kc/\sqrt{\epsilon_\infty}$ and the lower branch is constrained by the lines $\omega = kc/\sqrt{\epsilon_0}$ and $\omega = \omega_{\mathrm{TO}}$. For real values of ϵ_r, the reflectivity r of the crystal is given by

$$r = \left(\frac{\sqrt{\epsilon_r} - 1}{\sqrt{\epsilon_r} + 1}\right)^2. \tag{3.17}$$

Since $\epsilon_r(\omega_{\mathrm{TO}}) \to \infty$ we expect the reflection to approach unity at $\omega = \omega_{\mathrm{TO}}$, resulting in a steep step in the reflection spectra at frequencies just below ω_{TO}. This feature occurs at the edge of the *Reststrahlen band*, the spectral region for which $\epsilon_r < 0$ [25].

When an ionic crystal is placed in a finite electric field the polarisation of the atoms in the lattice changes [50]. This modifies the permittivity of the crystal; extensive theoretical work has sought in particular to predict the effect of finite electric fields on ϵ_∞ [50–52]. Since the phonon frequencies are determined by the dielectric function (Eqs. 3.15 and 3.16) applying an electric field shifts the Reststrahlen band. This effect is exploited for mid-infrared opto-electronic modulation applications in Chap. 5.

hBN's layered crystal structure means that it has different dielectric functions for light incident along the ordinary and extraordinary axes of the crystal. In hBN the longitudinal and transverse order phonon wavelengths are 12.12 μm and 13.16 μm respectively along the extraordinary axis of the crystal (the type I lower Reststrahlen band) and 6.20 and 7.35 μm along its ordinary axis (the type II upper Reststrahlen band [53]. hBN has attracted a lot of attention because the permittivities along each axis can be of opposite signs at the same wavelength, a property characteristic of hyperbolic metamaterials [54, 55].

3.4 Fabrication of 2D Materials

As with the fabrication of plasmonic nanostructures (Sect. 2.3.2), one can use either top-down of bottom-up approaches when fabricating 2D materials. Top-down approaches involve the exfoliation of the desired 2D materials from a bulk material. Commonly used exfoliation techniques include mechanical exfoliation (sometimes called mechanical cleavage) [6] and liquid phase exfoliation (LPE) [56]. Bottom-up approaches grow the materials layer by layer. The most commonplace method for graphene growth is chemical vapour deposition (CVD) [57].

We shall now discuss the two methods of 2D material fabrication used for the work presented in this thesis: mechanical exfoliation of graphene and hBN, and CVD growth of graphene.

3.4.1 Mechanical Exfoliation

Mechanically exfoliated graphene is still in widespread use for fundamental 2D material research and the development of proof-of-concept devices. Although the technique is unfeasible for large-scale industrial use due to its low yield and throughput it remains popular because it still provides the highest-quality 2D crystals.

Graphene is exfoliated from highly ordered pyrolytic graphite (HOPG)[5] and hBN is exfoliated from artificially synthesised crystals with a typical size of several cubic millimetres made from powdered hBN in a nickel base solution [58]. First, low-tack tape is pressed against the bulk crystal and then peeled off. The same piece of tape is then repeatedly peeled back on itself to reduce the thickness of each crystal flake. However, this also reduces the average size of each crystal. When large-area crystals are required the tape is not re-peeled.

A substrate is then prepared by O_2/Ar plasma etching to clean the surface and promote adhesion between the crystals and substrate. Immediately after plasma etching, the tape with exfoliated 2D crystals is pressed on to the substrate. We use Si/SiO_2 wafers as the substrate. The SiO_2 layer creates interference effects which enhances the optical contrast of 2D crystals and allows for easier identification of atomically thin regions [59]. Although the best contrast is achieved with a SiO_2 thickness of 90 nm, exfoliation is usually done on substrates with SiO_2 thicknesses of 290 nm, which is more favourable for studies where electrostatic gating is required.

Figure 3.7 shows an example of a mechanically exfoliated graphene flake. It is easy to distinguish between flake regions which are single-layer, bilayer and few-layer. Note that this flake is at the larger end of what one typically produces with mechanical exfoliation; single-layer graphene regions isolated using mechanical exfoliation are typically a few tens of microns across. The size of flake selected has been influenced by the required geometry of the final device. In this thesis we have characterised our devices using optical techniques; in particular, measuring reflection spectra using our spectroscopic ellipsometer at high angles of incidence necessitates the use of flakes with a size of around 100×50 m^2.

3.4.1.1 Wet Transfer of Exfoliated Two-Dimensional Materials

Three widely used techniques exist to transfer mechanically exfoliated 2D crystals from their initial Si/SiO_2 substrate to the target device (in the case of Chap. 5, this was a plasmonic nanostripe array). These techniques are commonly known as wet transfer, dry transfer and stamp transfer. We have chosen to use wet transfer technique since this tends to provide flakes with the largest areas.

First, a layer of 3% PMMA 950 K is spin-coated on the substrate with exfoliated 2D crystals (spin speed 5,000 rpm for 60 s, then baked on a hot plate at 130 °C for 3 min). A tape window (a 4×4 mm^2 piece of tape with a hole diameter 2 mm punched in the centre) is then attached on top of the PMMA film so that the desired 2D crystal flake is approximately in the centre of the tape window. A pair of sharp

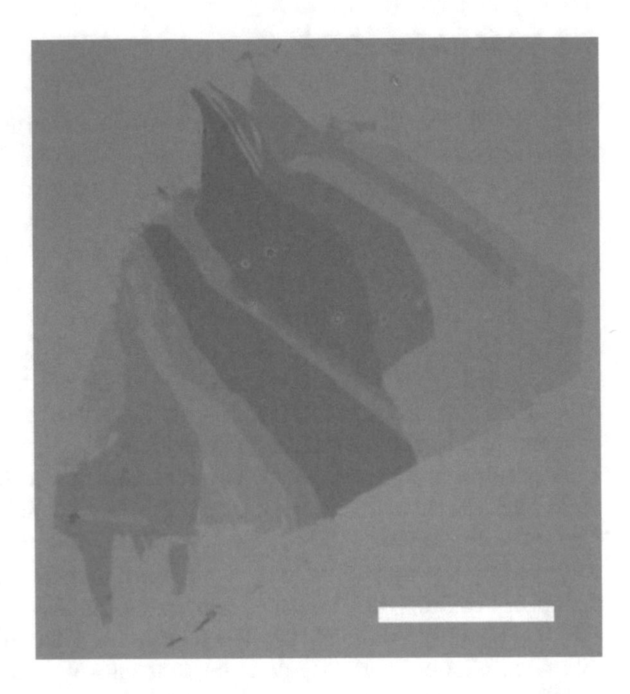

Fig. 3.7 Regions of mechanically exfoliated single-layer, bilayer and few-layer graphene. Also visible are bubbles characteristic of the mechanical exfoliation technique. Scale bar 100 m. Picture and sample by Francisco J. Rodriguez, used in Chap. 5

tweezers is used to cut through the PMMA film around the edge of the tape so that it is no longer attached to the rest of the PMMA on the substrate. The sample is then left in a KOH solution (3% weight by volume). The KOH etches the SiO_2 on the substrate, allowing the window to float up to the surface of the solution; this takes between 3 and 12 h. The window is then transferred to a deionised (DI) water solution for 10 min to rinse away the KOH solution from the underside of the window.

This completes the preparation of the window which is then ready to be transferred to the target substrate. We achieve this using a transfer rig. The transfer rig consists of a transfer arm which holds the window above the target substrate. The position of the window on the transfer arm is controlled with an x-y-z micromanipulator and the substrate can be rotated on a vacuum stage. Both the transfer arm and vacuum stage are directly below the objective of a microscope to check the alignment of the flake and substrate. The transfer arm is gradually moved down into contact with the substrate. Both the vacuum stage and transfer arm are heated: their temperatures are increased to around 70 °C once the window and substrate are in contact to promote adhesion. After a few minutes the PMMA window is lifted up and the 2D crystal should be left on the substrate. The sample is then baked on a hot plate at 130 °C for 10 min to improve adhesion further. Finally, the sample is cleaned in acetone and IPA to remove any residual PMMA before moving on to any subsequent processing.

The wet transfer technique does limit quality of graphene due to greater levels of contamination from being left in solution for an extended length of time. The dry transfer technique does not require the sample to be left in solution and so provides higher-quality graphene flakes, although contamination still exists from contact with the PMMA film [60]. The stamp transfer technique is used to encapsulate graphene and allows for the highest-quality graphene samples since the graphene flake never comes into contact with PMMA (instead being picked up by hBN while it is attached to a PMMA film) [61]. In our case, however, we are not interested in making high-quality electron transport measurements. Crucially, neither the dry nor stamp transfer processes are well-suited to the transfer of as large flakes as the wet transfer method.

3.4.2 *Growth of Graphene via Chemical Vapour Deposition*

Larger-scale areas of graphene can be fabricated using CVD growth. Early attempts to grow graphene used nickel films as a substrate; however, the size of graphene grown on these films was limited to the size of the grains on the Ni film [62]. Since the lateral size of one of these grains is around 20 m this did not represent an improvement over mechanically exfoliated flakes. It was later found that using copper foils as the substrate allowed for much larger continuous areas of graphene because annealing the copper gives much larger grain sizes. CVD growth of graphene on copper was initially reported to yield continuous regions of graphene with a lateral size of a few centimetres [57]; the technique was quickly developed to allow for industrial-scale graphene film production [63].

Typically, copper foils a few tens of microns in thickness are used. To remove the native oxide layer, the copper foil is annealed in a H_2 atmosphere at 1000 °C. A flow of H_2 and methane is then introduced at the same temperature at a pressure of around 500 mTorr. Graphene films are then formed on the foil *via* a surface-catalysed process within an hour [57, 64].

CVD graphene films tend to have a lower electron mobility than mechanically exfoliated films. Small islands of bilayer graphene are also very likely to appear in CVD graphene samples which may further affect their electronic performance and is the primary reason for the continued use of exfoliated graphene in fundamental and proof-of-concept studies. However, since our work with CVD graphene in Chap. 7 primarily uses graphene for protection against corrosion this is of little concern to us.

3.4.2.1 Transfer of CVD Graphene

All CVD graphene used in this thesis was provided by BGT Materials and Graphenea. The graphene covered both sides of a copper foil and was shipped in A4 (BGT) or square inch-sized pieces (Graphenea). To transfer CVD graphene to a target substrate, one must cover the graphene with a supporting polymer film, etch away the copper foil, transfer the graphene to the target substrate and then dissolve the polymer film. This process is described in detail below and illustrated in Fig. 3.8.

First, a piece of graphene on copper foil approximately 3×3 cm^2 in size was cut using scissors. A film of PMMA 950 K was spin-coated (3,000 rpm for 60 s) on one side of the foil. In general PMMA with a concentration ranging from 3 to 8% in

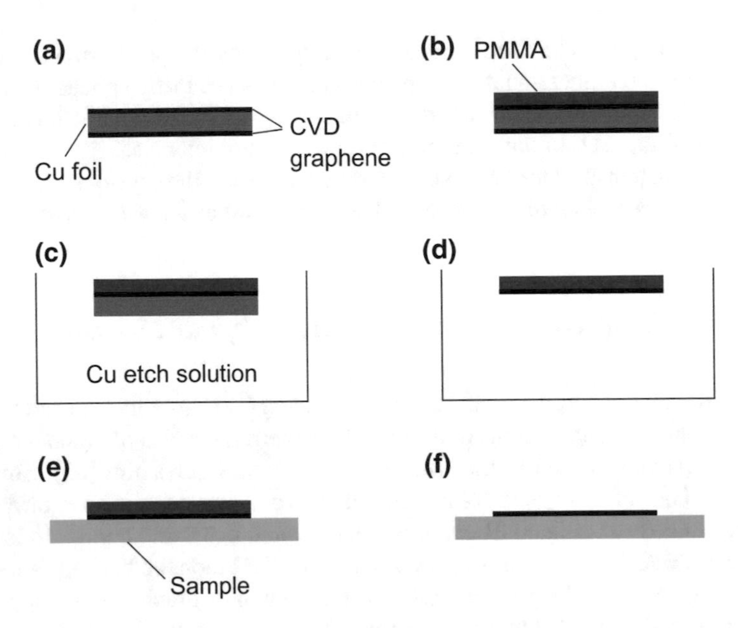

Fig. 3.8 Illustration of CVD transfer method. **a** CVD graphene grown on both sides of a copper foil. **b** Spin-coating of PMMA layer on one side of foil. **c** After plasma etch of one side of foil to remove graphene and expose copper, leave in copper etch solution overnight. **d** Copper foil dissolves in solution. **e** Rinse sample in DI water and transfer to sample. **f** After annealing to improve adhesion, dissolve PMMA layer in acetone and wash in IPA and hexane

anisol is used. In principle a lower concentration PMMA film will be more flexible than a higher concentration PMMA film, allowing for better adhesion to the target substrate. However, this also makes the PMMA and graphene more fragile during the transfer process. In this work 8% concentration PMMA was used.

Graphene is removed from the uncoated side of the copper film by O_2/Ar plasma etching for 60 s. The foil is cut into pieces of the size desired for the final samples (typically around 1×1 cm^2). The required samples are then placed in a solution of copper etchant (10–15 g/L ammonium persulphate ($(NH_4)_2S_2O_8$)) with the graphene and PMMA facing up and left overnight. Once the copper has been fully etched each sample is transferred to a beaker of deionised water for approximately five minutes to wash away the copper etchant. For the work presented in Chap. 7, where the target substrate is a plasmonic copper thin film, the samples are transferred to three beakers of deionised water for approximately five minutes each to ensure all copper etchant has been washed away. Samples are moved from one solution to the next by being picked up using a silicon wafer (chosen for its smoothness), taking care to ensure the films are not excessively flexed in the process.

The graphene/PMMA is transferred to the sample by picking the graphene out of the water using the final target substrate. The sample is then left to dry: water will naturally be driven from underneath the graphene by capillary forces. The processes

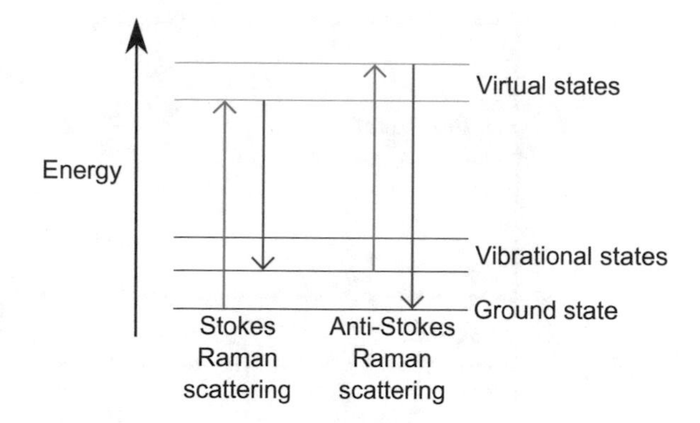

Fig. 3.9 Illustration of Stokes and Anti-Stokes Raman scattering modes

can be sped up by drying with a nitrogen gun for 2–3 min. To promote adhesion the sample is then heated. In most cases this is achieved by baking the sample on a hot plate at 150 °C for 10 min. In the case where heating in a standard atmosphere would damage the sample (as with our plasmonic copper films), the sample is instead annealed in a H_2/Ar atmosphere at 150 °C for 3–4 h.

Finally, the PMMA film is removed. The sample is placed in a beaker of acetone for 5 min, fresh acetone for another 5 min, IPA for 5 min and hexane for 1 min, then left to dry for approximately 1 min. The sample is inspected using an optical microscope to ascertain the quality of the graphene transfer.

3.5 Characterisation of 2D Materials

3.5.1 Raman Spectroscopy

Raman scattering was first demonstrated in 1928 by Raman and Krishnan [65]. They argued that if Compton scattering allows for the inelastic scattering of X-rays by electrons an analogous effect should occur for light inelastically scattered by atoms or molecules.

Two fundamental Raman scattering modes exist, illustrated in Fig. 3.9. In Stokes scattering a photon excites an electron in an atom or molecule from its ground state to a virtual energy level. The electron then returns to the first vibrational excitation state. The first vibrational excitation state has a higher energy than the ground state, meaning that the scattered photon has a lower energy than the incident photon. In the case of anti-Stokes scattering the electron starts in the first vibrational energy level, is excited to a higher virtual energy level by photon absorption and then returns to its

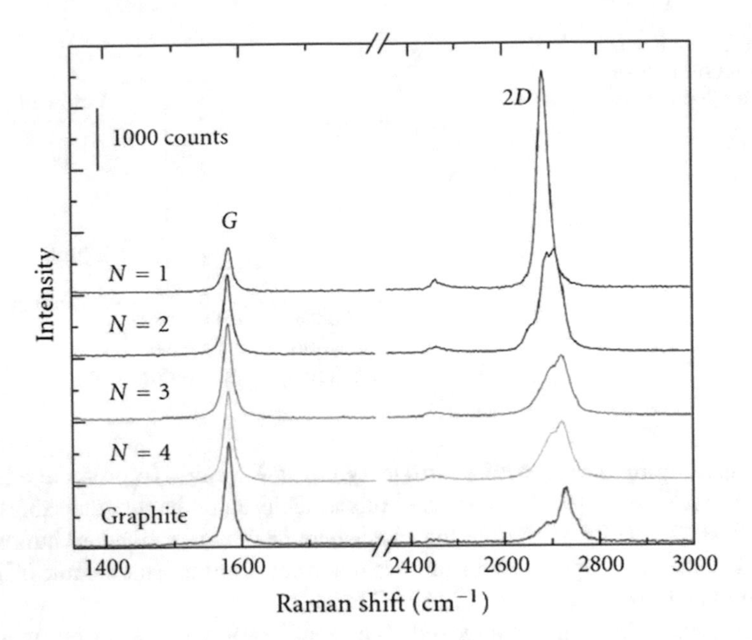

Fig. 3.10 Raman spectrum of graphene showing the evolution of the G and 2D peaks as a function of the number of graphene layers. Reproduced from [26]

ground state, meaning that the scattered photon has a higher energy than the incident photon.[1]

Raman spectroscopy is used to identify molecules in a sample. A sample is irradiated with laser light and any shifts in the frequency of light scattered from the sample due to Raman scattering are measured. Each molecule will have its own distinctive set of shifts, usually plotted in a Raman spectrum as relative intensity of certain peaks (corresponding to specific transitions) plotted as a function of wavenumber (in units of cm^{-1}).

Graphene's phonon dispersion relation consists of 3 optical phonon branches and 3 acoustic phonon branches [66]. One optical and one acoustic branch give atomic vibrations perpendicular to the graphene sheet (out-of-plane phonon modes) while two optical and two acoustic branches give rise to in-plane atomic vibrations. Vibrational modes are classified as either transverse or longitudinal depending on whether of not they are parallel or perpendicular to the A-B carbon bond. Not all of these vibrational modes are Raman active [33].

Figure 3.10 shows the Raman spectrum of single- and few-layer graphene. The G peak at $\sim$1580 cm^{-1} arises from doubly degenerate in-plane transverse and

[1] For completeness we also note the existence of two non-Raman scattering processes. In Rayleigh scattering the electron is excited from and returns to the ground state, meaning that the incident and scattered photons have the same wavelength. In fluorescence a photon excites an electron to a higher energy state. The electron returns to the ground state via a series of non-radiative transitions followed by one large transition. The final transition corresponds to fluorescent photon emission.

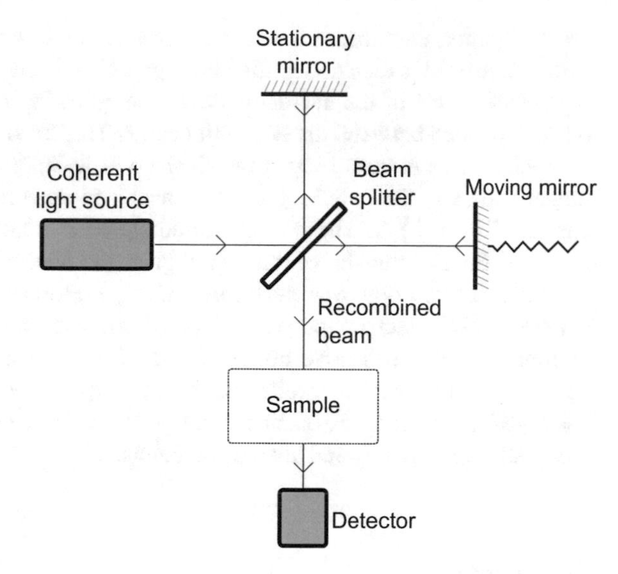

Fig. 3.11 Operating principle of a Fourier transform infrared spectrometer

longitudinal optical phonon modes near the Brillouin zone centre Γ (which belong to the 2D E_{2g} representation according to group theory [66]).

A D peak at ~1350 cm^{-1} (not shown in Fig. 3.10) is due to the "breathing mode" of the six-atom ring (where phonon modes point out of the ring). It is double-resonant and requires a defect to be activated [33, 67]. In defect-free graphene it can only be seen on the edge of a graphene flake [68]. The D' peak is a double resonance connecting two points in the same cone around the K or K' point and lies on the upper shoulder of the G peak [33]. The 2D (~2700 cm^{-1}) and 2D' (~3250 cm^{-1}) peaks are overtones of the D and D' peaks, respectively, which do not require defects to be activated since momentum conservation is satisfied by two phonons with opposite wavevectors. The 2D peak is historically referred to as G' [68].

Raman spectroscopy is an integral part of graphene research [33]. Figure 3.10 demonstrates the significant changes in the 2D peak for graphene flakes of various thicknesses. This change in shape and intensity allows for easy verification of the number of layers of graphene in a flake [68].

3.5.2 Fourier Transform Infrared Spectroscopy

Plasmonic features usually appear in visible-NIR region and are best characterised using spectroscopic ellipsometry or direct reflection/transmission spectroscopy. However, for optical features that lie in the near-mid infrared region Fourier transform infrared (FTIR) spectroscopy is used.

Figure 3.11 shows the basic schematic of an FTIR spectrometer. A collimated beam of light is passed into a Michelson interferometer. The light passes through a beam splitter, each beam is reflected from a mirror and recombines at the beam splitter. The light is then transmitted through or is reflected from the sample and goes to a detector. One of the mirrors is moveable, allowing for a path length difference to be introduced between the two split beams. This means that when the two beams recombine there is constructive interference at wavelengths of $n\Delta$ (where n is an integer and Δ is the path length difference) and destructive interference at wavelengths of $\left(n + \frac{1}{2}\right)\Delta$. The raw measured signal is relative intensity as a function of Δ. The Fourier transfer of this signal gives the final FTIR spectrum.

FTIR spectroscopy was performed using a Bruker Vertex 80 system with a Hyperion 3000 microscope. A variety of sources and detectors combined with aluminium-coated reflective optics enabled this system to be used from the visible to THz wavelengths. A reflecting objective provides a range of incident angles ($\theta = 12$–$24°$) and the entire beam path purged with dry, CO_2-scrubbed air to minimise strong atmospheric infrared absorption bands.

References

1. P.R. Wallace, The band theory of graphite. Phys. Rev. **71**(9), 622 (1947)
2. R. Peierls, Quelques propriétés typiques des corps solides. Annales de l'institut Henri Poincaré **5**, 177–222 (1935)
3. L.D. Landau, Zur theorie der phasenumwandlungen ii. Phys. Z. Sowjetunion **11**, 26–35 (1937)
4. A.K. Geim, K.S. Novoselov, The rise of graphene. Nat. Mater. **6**(3), 183–191 (2007)
5. K.S. Novoselov, A.K. Geim, S.V. Morozov, D. Jiang, Y. Zhang, S.V. Dubonos, I.V. Grigorieva, A.A. Firsov, Electric field effect in atomically thin carbon films. Science **306**(5696), 666–669 (2004)
6. K.S. Novoselov, A.K. Geim, S.V. Morozov, D. Jiang, M.I. Katsnelson, I.V. Grigorieva, S.V. Dubonos, A.A. Firsov, Two-dimensional gas of massless dirac fermions in graphene. Nature **438**(7065), 197–200 (2005)
7. R.R. Nair, P. Blake, A.N. Grigorenko, K.S. Novoselov, T.J. Booth, T. Stauber, N.M.R. Peres, A.K. Geim, Fine structure constant defines visual transparency of graphene. Science **320**(5881), 1308–1308 (2008)
8. E.V. Castro, H. Ochoa, M.I. Katsnelson, R.V. Gorbachev, D.C. Elias, K.S. Novoselov, A.K. Geim, F. Guinea, Limits on charge carrier mobility in suspended graphene due to flexural phonons. Phys. Rev. Lett. **105**(26), 266601 (2010)
9. F. Wang, Y. Zhang, C. Tian, C. Girit, A. Zettl, M. Crommie, Y.R. Shen, Gate-variable optical transitions in graphene. Science **320**(5873), 206–209 (2008)
10. C. Lee, X. Wei, J.W. Kysar, J. Hone, Measurement of the elastic properties and intrinsic strength of monolayer graphene. Science **321**(5887), 385–388 (2008)
11. A.A. Balandin, S. Ghosh, W. Bao, I. Calizo, D. Teweldebrhan, F. Miao, C.N. Lau, Superior thermal conductivity of single-layer graphene. Nano Lett. **8**(3), 902–907 (2008)
12. C. Liu, Z. Yu, D. Neff, A. Zhamu, B.Z. Jang, Graphene-based supercapacitor with an ultrahigh energy density. Nano Lett. **10**(12), 4863–4868 (2010)
13. F. Bonaccorso, Z. Sun, T. Hasan, A.C. Ferrari, Graphene photonics and optoelectronics. Nat. Photon. **4**(9), 611–622 (2010)
14. S. Stankovich, D.A. Dikin, G.H.B. Dommett, K.M. Kohlhaas, E.J. Zimney, E.A. Stach, R.D. Piner, S.T. Nguyen, R.S. Ruoff, Graphene-based composite materials. Nature **442**(7100), 282–286 (2006)
15. M.D. Stoller, S. Park, Y. Zhu, J. An, R.S. Ruoff, Graphene-based ultracapacitors. Nano Lett. **8**(10), 3498–3502 (2008)

16. C. Chung, Y.-K. Kim, D. Shin, S.-R. Ryoo, B.H. Hong, D.-H. Min, Biomedical applications of graphene and graphene oxide. Acc. Chem. Res. **46**(10), 2211–2224 (2013)
17. M. Xu, T. Liang, M. Shi, H. Chen, Graphene-like two-dimensional materials. Chem. Rev. **113**(5), 3766–3798 (2013)
18. A.K. Geim, I.V. Grigorieva, Van der waals heterostructures. Nature **499**(7459), 419–425 (2013)
19. C.R. Dean, A.F. Young, I. Meric, C. Lee, L. Wang, S. Sorgenfrei, K. Watanabe, T. Taniguchi, P. Kim, K.L. Shepard, J. Hone, Boron nitride substrates for high-quality graphene electronics. Nat. Nanotechnol. **5**(10), 722–726 (2010)
20. R.V. Gorbachev, J.C.W. Song, G. Yu, A.V. Kretinin, F. Withers, Y. Cao, A. Mishchenko, I.V. Grigorieva, K.S. Novoselov, L.S. Levitov, A.K. Geim, Detecting topological currents in graphene superlattices. Science **346**(6208), 448–451 (2014)
21. D.A. Bandurin, I. Torre, R.K. Kumar, M.B. Shalom, A. Tomadin, A. Principi, G.H. Auton, E. Khestanova, K.S. Novoselov, I.V. Grigorieva, L.A. Ponomarenko, A.K. Geim, M. Polini, Negative local resistance caused by viscous electron backflow in graphene. Science **351**(6277), 1055–1058 (2016)
22. M.B. Shalom, M.J. Zhu, V.I. Falko, A. Mishchenko, A.V. Kretinin, K.S. Novoselov, C.R. Woods, K. Watanabe, T. Taniguchi, A.K. Geim, J.R. Prance, Quantum oscillations of the critical current and high-field superconducting proximity in ballistic graphene. Nat. Phys. **12**(4), 318–322 (2016)
23. M.I. Katsnelson, *Graphene: Carbon in Two Dimensions* (Cambridge University Press, Cambridge, 2012)
24. A.H.C. Neto, F. Guinea, N.M.R. Peres, K.S. Novoselov, A.K. Geim, The electronic properties of graphene. Rev. Mod. Phys. **81**(1), 109 (2009)
25. N.W. Ashcroft, N.D. Mermin, *Solid State Physics* (Harcourt Brace, Orlando, 1976)
26. D.R. Cooper, B. D'Anjou, N. Ghattamaneni, B. Harack, M. Hilke, A. Horth, N. Majlis, M. Massicotte, L. Vandsburger, E. Whiteway, V. Yu, Experimental review of graphene. ISRN Condens. Matter Phys. **2012**, 501686 (2012)
27. F. Schwierz, Graphene transistors. Nat. Nanotechnol. **5**(7), 487–496 (2010)
28. L.A. Falkovsky, A.A. Varlamov, Space-time dispersion of graphene conductivity. Eur. Phys. J. B-Condens. Matter Complex Syst. **56**(4), 281–284 (2007)
29. A.B. Kuzmenko, E. Van Heumen, F. Carbone, D. Van Der Marel, Universal optical conductance of graphite. Phys. Rev. Lett. **100**(11), 117401 (2008)
30. V.G. Kravets, A.N. Grigorenko, R.R. Nair, P. Blake, S. Anissimova, K.S. Novoselov, A.K. Geim, Spectroscopic ellipsometry of graphene and an exciton-shifted van hove peak in absorption. Phys. Rev. B **81**(15), 155413 (2010)
31. V.G. Kravets, F. Schedin, A.N. Grigorenko, Fine structure constant and quantized optical transparency of plasmonic nanoarrays. Nat. Commun. **3**, 640 (2012)
32. L. Van Hove, The occurrence of singularities in the elastic frequency distribution of a crystal. Phys. Rev. **89**(6), 1189 (1953)
33. A.C. Ferrari, D.M. Basko, Raman spectroscopy as a versatile tool for studying the properties of graphene. Nat. Nanotechnol. **8**(4), 235–246 (2013)
34. L. Yang, J. Deslippe, C.-H. Park, M.L. Cohen, S.G. Louie, Excitonic effects on the optical response of graphene and bilayer graphene. Phys. Rev. Lett. **103**(18), 186802 (2009)
35. L.A. Falkovsky, Optical properties of graphene, in *Journal of Physics: Conference Series*, vol. 129 (IOP Publishing, 2008), p. 012004
36. Z.Q. Li, E.A. Henriksen, Z. Jiang, Z. Hao, M.C. Martin, P. Kim, H.L. Stormer, D.N. Basov, Dirac charge dynamics in graphene by infrared spectroscopy. Nat. Phys. **4**(7), 532–535 (2008)
37. Z. Sun, A. Martinez, F. Wang, Optical modulators with 2d layered materials. Nat. Photon. **10**(4), 227–238 (2016)
38. F.J. Garcia de Abajo, Graphene plasmonics: challenges and opportunities. ACS Photon. **1**(3), 135–152 (2014)
39. F.H.L. Koppens, D.E. Chang, F.J. García de Abajo, Graphene plasmonics: a platform for strong light-matter interactions. Nano Lett. **11**(8), 3370–3377 (2011)

40. H. Yan, T. Low, W. Zhu, Y. Wu, M. Freitag, X. Li, F. Guinea, P. Avouris, F. Xia, Damping pathways of mid-infrared plasmons in graphene nanostructures. Nat. Photon. **7**(5), 394–399 (2013)
41. A. Woessner, M.B. Lundeberg, Y. Gao, A. Principi, P. Alonso-González, M. Carrega, K. Watanabe, T. Taniguchi, G. Vignale, M. Polini, J. Hone, R. Hillenbrand, F.H.L. Koppens, Highly confined low-loss plasmons in graphene-boron nitride heterostructures. Nat. Mater. **14**(4), 421–425 (2015)
42. T. Low, P. Avouris, Graphene plasmonics for terahertz to mid-infrared applications. ACS Nano **8**(2), 1086–1101 (2014)
43. R.H. Wentorf, Cubic form of boron nitride. J. Chem. Phys. **26**(4), 956–956 (1957)
44. T. Sōma, A. Sawaoka, S. Saito, Characterization of wurtzite type boron nitride synthesized by shock compression. Mater. Res. Bull. **9**(6), 755–762 (1974)
45. R.S. Pease, An x-ray study of boron nitride. Acta Crystallogr. **5**(3), 356–361 (1952)
46. G. Giovannetti, P.A. Khomyakov, G. Brocks, P.J. Kelly, J. Van Den Brink, Substrate-induced band gap in graphene on hexagonal boron nitride: Ab initio density functional calculations. Phys. Rev. B **76**(7), 073103 (2007)
47. K. Watanabe, T. Taniguchi, H. Kanda, Direct-bandgap properties and evidence for ultraviolet lasing of hexagonal boron nitride single crystal. Nat. Mater. **3**(6), 404–409 (2004)
48. L. Song, L. Ci, H. Lu, P.B. Sorokin, C. Jin, J. Ni, A.G. Kvashnin, D.G. Kvashnin, J. Lou, B.I. Yakobson, A.M. Pulickel, Large scale growth and characterization of atomic hexagonal boron nitride layers. Nano Lett. **10**(8), 3209–3215 (2010)
49. Y.-N. Xu, W.Y. Ching, Calculation of ground-state and optical properties of boron nitrides in the hexagonal, cubic, and wurtzite structures. Phys. Rev. B **44**(15), 7787 (1991)
50. P. Umari, A. Pasquarello, Ab initio molecular dynamics in a finite homogeneous electric field. Phys. Rev. Lett. **89**(15), 157602 (2002)
51. X. Wang, D. Vanderbilt, First-principles perturbative computation of phonon properties of insulators in finite electric fields. Phys. Rev. B **74**(5), 054304 (2006)
52. X. Wang, D. Vanderbilt, First-principles perturbative computation of dielectric and born charge tensors in finite electric fields. Phys. Rev. B **75**(11), 115116 (2007)
53. J.D. Caldwell, L. Lindsay, V. Giannini, I. Vurgaftman, T.L. Reinecke, S.A. Maier, O.J. Glembocki, Low-loss, infrared and terahertz nanophotonics using surface phonon polaritons. Nanophotonics **4**(1), 44–68 (2015)
54. J.D. Caldwell, A.V. Kretinin, Y. Chen, V. Giannini, M.M. Fogler, Y. Francescato, C.T. Ellis, J.G. Tischler, C.R. Woods, A.J. Giles, M. Hong, K. Watanabe, T. Taniguchi, S.A. Maier, K.S. Novoselov, Sub-diffractional volume-confined polaritons in the natural hyperbolic material hexagonal boron nitride. Nat. Commun. **5**, 5221 (2014)
55. S. Dai, Q. Ma, T. Andersen, A.S. Mcleod, Z. Fei, M.K. Liu, M. Wagner, K. Watanabe, T. Taniguchi, M. Thiemens, F. Keilmann, P. Jarillo-Herrero, M.M. Fogler, D.N. Basov, Sub-diffractional focusing and guiding of polaritonic rays in a natural hyperbolic material. Nat. Commun. **6** (2015)
56. Y. Hernandez, V. Nicolosi, M. Lotya, F.M. Blighe, Z. Sun, S. De, I.T. McGovern, B. Holland, M. Byrne, Y.K. Gun'Ko, J.J. Boland, P. Nirag, G. Duesberg, S. Krishnamurthy, R. Goodhue, J. Hutchinson, V. Scardaci, A.C. Ferrari, J.N. Coleman, High-yield production of graphene by liquid-phase exfoliation of graphite. Nat. Nanotechnol. **3**(9), 563–568 (2008)
57. X. Li, W. Cai, J. An, S. Kim, J. Nah, D. Yang, R. Piner, A. Velamakanni, I. Jung, E. Tutuc, S.K. Banerjee, L. Colombo, R.S. Ruoff, Large-area synthesis of high-quality and uniform graphene films on copper foils. Science **324**(5932), 1312–1314 (2009)
58. Y. Kubota, K. Watanabe, O. Tsuda, T. Taniguchi, Deep ultraviolet light-emitting hexagonal boron nitride synthesized at atmospheric pressure. Science **317**(5840), 932–934 (2007)
59. P. Blake, E.W. Hill, A.H.C. Neto, K.S. Novoselov, D. Jiang, R. Yang, T.J. Booth, A.K. Geim, Making graphene visible. Appl. Phys. Lett. **91**(6), 063124 (2007)
60. A.V. Kretinin, Y. Cao, J.S. Tu, G.L. Yu, R. Jalil, K.S. Novoselov, S.J. Haigh, A. Gholinia, A. Mishchenko, M. Lozada, T. Georgiou, C.R. Woods, F. Withers, P. Blake, G. Eda, A. Wirsig, C. Hucho, K. Watanabe, T. Taniguchi, A.K. Geim, R.V. Gorbachev, Electronic properties of

graphene encapsulated with different two-dimensional atomic crystals. Nano Lett. **14**(6), 3270–3276 (2014)
61. L. Wang, I. Meric, P.Y. Huang, Q. Gao, Y. Gao, H. Tran, T. Taniguchi, K. Watanabe, L.M. Campos, D.A. Muller, J. Guo, P. Kim, J. Hone, K.L. Shepard, C.R. Dean, One-dimensional electrical contact to a two-dimensional material. Science **342**(6158), 614–617 (2013)
62. A. Reina, X. Jia, J. Ho, D. Nezich, H. Son, V. Bulovic, M.S. Dresselhaus, J. Kong, Large area, few-layer graphene films on arbitrary substrates by chemical vapor deposition. Nano Lett. **9**(1), 30–35 (2008)
63. S. Bae, H. Kim, Y. Lee, X. Xu, J.-S. Park, Y. Zheng, J. Balakrishnan, T. Lei, H.R. Kim, Y.I. Song, Y.-J. Kim, K.S. Kim, B. Özyilmaz, B.H. Ahn, J.-H. Hong, S. Iijima, Roll-to-roll production of 30-inch graphene films for transparent electrodes. Nat. Nanotechnol. **5**(8), 574–578 (2010)
64. V.G. Kravets, R. Jalil, Y.-J. Kim, D. Ansell, D.E. Aznakayeva, B. Thackray, L. Britnell, B.D. Belle, F. Withers, I.P. Radko, Z. Han, S.I. Bozhevolnyi, K.S. Novoselov, A.K. Geim, A.N. Grigorenko, Graphene-protected copper and silver plasmonics. Sci. Rep. **4** (2014)
65. C.V. Raman, K.S. Krishnan, A new type of secondary radiation. Nature **121**, 501–502 (1928)
66. L.M. Malard, M.A.A. Pimenta, G. Dresselhaus, M.S. Dresselhaus, Raman spectroscopy in graphene. Phys. Rep. **473**(5), 51–87 (2009)
67. C. Thomsen, S. Reich, Double resonant raman scattering in graphite. Phys. Rev. Lett. **85**(24), 5214 (2000)
68. A.C. Ferrari, J.C. Meyer, V. Scardaci, C. Casiraghi, M. Lazzeri, F. Mauri, S. Piscanec, D. Jiang, K.S. Novoselov, S. Roth, A.K. Geim, Raman spectrum of graphene and graphene layers. Phys. Rev. Lett. **97**(18), 187401 (2006)

Chapter 4
Super-Narrow, Extremely High Quality Collective Plasmon Resonances at Telecommunication Wavelengths

In this chapter we describe the theory, fabrication and characterisation of gold nanostripe arrays on a thin gold film, with the spectral line full width at half-maximum (FWHM) as low as 5 nm and quality factors Q reaching 300, at important fibre-optic telecommunication wavelengths around 1.5 μm. My work builds on preliminary results of Benjamin Thackray presented in Chap. 3 of his thesis [1]: I optimised the nanoarrays to improve on his results and extended the study to analyse the effect of the height of the nanostructure. The results in this chapter were published in *Nano Letters* in 2015 [2].

I designed the samples based on an initial design by Benjamin Thackray. I fabricated the samples with assistance from Gregory Auton for electron-beam lithography and Vasyl Kravets for sample preparation and lift-off. I characterised samples using ellipsometry. Scanning Electron Microscopy characterisation was performed by Benjamin Thackray and Francisco Rodriguez. I coauthored the article manuscript with Benjamin Thackray and Alexander Grigorenko.

4.1 Introduction

Plasmon resonances in metallic nanostructures have attracted a high level of interest in recent years for their promising applications spanning many fields, from negative index metamaterials and perfect lensing [3] to extremely sensitive biosensing [4, 5]. For key applications, such as sensing and active plasmonics, it is crucial to have the narrowest plasmon resonances possible.

As described in Sect. 2.2.3, small changes to the ambient medium around a plasmonic nanostructure will shift the position of its plasmon resonance. This shifting is not necessarily greater in magnitude for plasmonic nanostructures with narrower resonances, but changes in measured reflection amplitude can potentially be much more dramatic with narrow resonances. The edges of a narrow resonance have a higher

P. A. Thomas, *Narrow Plasmon Resonances in Hybrid Systems*,
Springer Theses, https://doi.org/10.1007/978-3-319-97526-9_4

gradient than the edges of a broad resonance. Therefore, the reflectance measured at a wavelength corresponding to a point on the edge of the resonance will change more dramatically for a narrow resonance.[1]

The sharpness of a resonance is quantified by its quality factor:

$$Q = \frac{\lambda_R}{\Delta\lambda}, \tag{4.1}$$

where λ_R is the wavelength of its resonance and $\Delta\lambda$ is its full width at half maximum (FWHM). Surface plasmon polaritons propagating in a continuous film generally provide a resonance quality factor at the level of $Q < 20$ [6]. Localised plasmon resonances observed in isolated nanoparticles tend to be even broader (typically $Q < 10$ [7]), limiting their use in biosensing and light modulation.

The spectral width of the resonance peak can be narrowed by coupling resonance modes in regular nanostructure arrays, providing ultranarrow collective, diffraction coupled plasmon resonances [8–11], also known as geometric resonances. If arrays are fabricated so that the Rayleigh diffraction anomalies of the array [12] (where light is diffracted parallel to the plane of the array as a diffraction mode crosses from the air into the substrate) and the localised surface plasmon resonance modes of the structures occur at similar wavelengths, then light that would otherwise be scattered to the far-field can be recaptured as electron oscillations in the neighbouring nanostructures in phase with plasmon excitation induced by the incident light. By using the right combination of nanostructure size, shape, and array period, one can achieve ultranarrow and deep collective plasmon resonances at the desired wavelength that normally improve with increasing array size and/or spatial coherence of the beam.

In this work, we achieve a significant improvement in the quality factor of the collective resonances designed for telecom wavelengths. We have measured some of the highest recorded values of Q for collective resonances in diffraction coupled arrays of plasmonic nanostructures, registering $Q \approx 300$ at wavelengths of around 1.5 μm. This improvement was achieved by adding a continuous gold layer beneath a gold nanostripe array (design shown in Fig. 4.2a) and utilizing a large angle of incidence (around 80°), which generates large image dipoles in the gold sublayer and mimics an index-matched environment. Similar results, albeit in the less important mid-infrared region (from 3 to 5 μm), were previously reported by Li et al. [13].

4.2 Diffraction Coupling of Localised Plasmon Resonances

The low quality of single nanoparticle plasmon resonances can be understood through the Frölich condition for the maximum polarisability of a nanoparticle ($\text{Re}\,[\epsilon_1] = -2\epsilon_2$; see Sect. 2.3.1). The metal nanoparticle's dielectric function ϵ_1 contains a non-

[1]For completeness, we note that an extra level of sensitivity can be achieved by measuring changes in phase instead of reflection for resonances with close to zero reflection at their maximum strength. This idea is explored in greater depth in Chap. 7.

negligible imaginary component while the ambient medium's dielectric constant ϵ_2 is usually near unity. Therefore it is usually impossible to perfectly satisfy the Frölich condition, leading to a spectrally broad resonance with relatively low absorption at its peak. The value of Q for a given material with a given plasmon resonance is independent of the size or shape of the geometry [14]. To increase Q for plasmonic nanoparticles we must therefore employ coupling effects.

A number of mechanisms exist to couple localised plasmon resonances, including resistive [15] and near-field [16] coupling effects. Near-field coupling occurs when the electron clouds from adjacent nanoparticles influence one another to created hybridised plasmon modes, including some modes which can allow for marginal increases in Q [17]. However, the most dramatic enhancements in Q have been achieved using far-field diffraction coupling of localised plasmon resonances.

When an external light source is incident upon an ordered nanoparticle array on a transparent substrate, grating-like diffraction will occur along each axis of the array. Some diffraction modes will run along the plane of the array. The wavelengths of these modes are predicted by the Rayleigh cutoff wavelength λ_R of Wood diffraction anomalies [10]. Transitions of diffracted modes from the air to the substrate (and vice versa) are forbidden because light has a different dispersion relationship in each media. Two sets of cutoff wavelengths exist: one corresponding to the disappearance of a diffraction mode crossing from air into the substrate λ_R^{air} (the "air" diffraction modes) and another to the disappearance of a diffraction crossing from the substrate into the air λ_R^{sub} (the "substrate" diffraction modes). The air diffraction modes for a regular square array with periodicity a are given by

$$\lambda_{R(m,p,q)}^{\text{air}} = \frac{a}{m}\left[\sqrt{p^2+q^2} \pm \sin(\theta)\left\langle p\cos(\phi) + q\sin(\phi)\right\rangle\right], \tag{4.2}$$

where m, p, and q are integers, θ is the angle of incidence, ϕ is the angle between the axis of the substrate and the plane of incidence (the substrate orientation) and we have taken the refractive index of air to be 1. For $\phi = 0$ the most important diffraction modes are given by

$$\lambda_R^{\text{air}} = \frac{a}{m}\left(n_a + \sin\theta\right) \tag{4.3}$$

$$\lambda_R^{\text{sub}} = \frac{a}{m}\left(n_s + \sin\theta\right), \tag{4.4}$$

where n_a is the refractive index of the ambient medium (usually air, with $n_a = 1$) and n_s is the refractive index of the substrate. These diffraction modes can be recaptured in plasmon resonance in the nanoparticles in the array. The size of the nanoparticles, period of the array and angle of incidence can be tuned so that the diffracted modes will be scattered into each nanoparticle in phase with the nanoparticle's plasmon resonance, thereby strengthening the plasmon resonance. Scattering can therefore act to negate the imaginary part of ϵ_1 and allow the Frölich condition to be satisfied. Additionally, the diffracted modes that would have been scattered to the far field are instead absorbed by the nanoparticles. Since

$$Q = \frac{f}{\Delta f} = \frac{\lambda}{\Delta \lambda} = 2\pi \frac{\text{Energy stored}}{\text{Energy dissipated}}, \tag{4.5}$$

this effect leads to an increase in Q. Therefore, in large arrays with hundreds of nanoparticles, diffraction coupling can result in extremely narrow plasmon resonances with strong absorption and local electric field enhancements.

4.2.1 Coupled Dipole Approximation

A rigorous mathematical description of diffraction coupling is given by the coupled dipole approximation [8, 9, 18, 19]. The coupled dipole approximation has its roots in works of DeVoe [20, 21] and Purcell and Pennypacker [22] published in the 1960s and 1970s. However, a well-developed mathematical description of this theory was not arrived upon until the works of Markel [18] and Schatz [8, 19] in the 1990s and 2000s.[2]

Our approach follows that of references [8, 23]. In the coupled dipole approximation we consider a one-dimensional array of N nanoparticles with positions $\mathbf{r}_i$ and polarisabilities α_i. The nanoparticle array is made up of identical particles arranged with a fixed periodicity a. We assume the particles are small compared to both wavelength of incident light and a. The nanoarray is illuminated by an external electric field $\mathbf{E}_{\text{inc}} = \mathbf{E}_0 e^{i(\mathbf{k}\cdot\mathbf{r}_i)}$ which induces a dipole in each particle given by

$$\mathbf{p}_i = \alpha_i \mathbf{E}_{\text{loc},i}, \tag{4.6}$$

where $\mathbf{E}_{\text{loc},i}$ is the local electric field at the position of the ith nanoparticle and $\mathbf{E}_{\text{loc},i}$ is the sum of $\mathbf{E}_{\text{inc}}$ and the retarded fields from the other $N-1$ particle dipoles, $\mathbf{E}_{\text{dip},i}$:

$$\mathbf{E}_{\text{loc},i} = \mathbf{E}_{\text{inc}} + \mathbf{E}_{\text{dip},i} = \mathbf{E}_0 e^{i(\mathbf{k}\cdot\mathbf{R}_i)} - \sum_{\substack{j=1 \\ j\neq i}}^{N} A_{ij}\cdot\mathbf{p}_j \qquad i = 1, 2, \ldots N. \tag{4.7}$$

A is the dipole interaction matrix, expressed as

$$A_{ij}\cdot\mathbf{p}_j = k^2 e^{ikr_{ij}} \frac{\mathbf{r}_{ij}\times\left(\mathbf{r}_{ij}\times\mathbf{p}_j\right)}{r_{ij}^3} + e^{ikr_{ij}}\left(1 - ikr_{ij}\right)\frac{\left[r_{ij}^2\mathbf{p}_j - 3\mathbf{r}_{ij}\left(\mathbf{r}_{ij}\cdot\mathbf{p}_j\right)\right]}{r_{ij}^5}, \tag{4.8}$$

where $\mathbf{r}_{ij} = \mathbf{r}_j - \mathbf{r}_i$ is the vector from dipole i to dipole j.

---

[2]For this reason the coupled dipole approximation is sometimes referred to as Markel-Schatz or Schatz-Markel theory [10].

The polarisation vectors can be obtained by solving $3N$ linear equations of the form

$$A'\mathbf{p} = \mathbf{E}. \tag{4.9}$$

An analytical solution exists when $N \rightarrow \infty$, $\mathbf{k} \perp \mathbf{r}_{ij}$ and $\mathbf{p}_i = \mathbf{p}$; that is, all particles in the array have the same induced dipole (and same polarisability $\alpha_i = \alpha_s$):

$$\mathbf{p} = \frac{\alpha_s E_0}{1 - \alpha_s S}. \tag{4.10}$$

S is the retarded dipole sum

$$S = \sum_{j \neq i} \left[\frac{(1 - ikr_{ij})(3\cos^2\theta_{ij} - 1)e^{ikr_{ij}}}{r_{ij}^3} + \frac{k^2 \sin^2\theta_{ij} e^{ikr_{ij}}}{r_{ij}} \right]. \tag{4.11}$$

θ_{ij} is the angle between $\mathbf{r}_{ij}$ and the direction of polarisation. We can rewrite Eq. 4.10 in terms of effective polarisability α_{eff}:

$$\alpha_{\text{eff}} = \frac{1}{\alpha_s^{-1} - S}. \tag{4.12}$$

α_{eff} will be maximal (and so the nanoarray will give the strongest resonances) when $\alpha_s^{-1} = S$. S depends on array period and angle of incidence; α_s depends on size of particle. Both quantities depend on the refractive indices of the particles, substrate and ambient medium. Therefore, the strongest resonances can be attained by tuning the array period and particle size, just as we discussed in the physical model of Rayleigh cutoff wavelengths.

The behaviour of the resonances for slight mismatches between the real parts of α_s^{-1} and S depend on which is larger [24]: for the case of $\text{Re}(S) > \text{Re}(\alpha_s^{-1})$ each diffraction-coupled resonance will split in two.

4.2.2 Previous Experimental Observations of Diffraction Coupled Plasmon Resonances

The first experimental studies that attempted to apply the coupled dipole approximation to the fabrication of plasmonic nanoarrays observed only a relatively small degree of narrowing of plasmon resonances [25–27]. These studies used focussing optics with numerical aperture (NA) >0.3, meaning their results were limited by spatial coherence. Theoretical studies of the coupled dipole approximation assume the dipoles in an array are excited by a perfectly collimated beam of light. In reality, however, the small size of the fabricated nanoarrays (no more than a few hundred

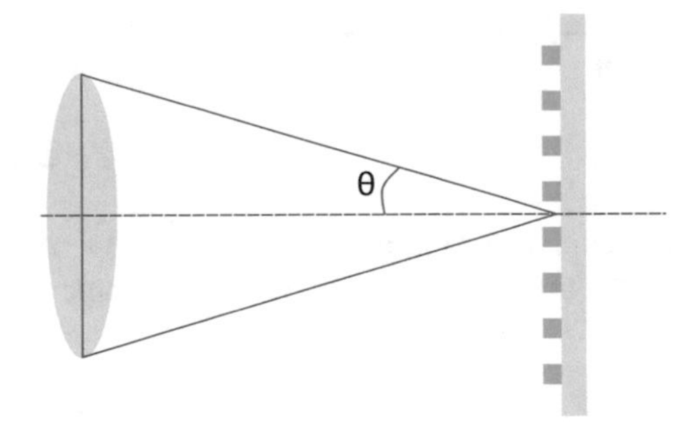

Fig. 4.1 Illustration of focussing optics. Spatial coherence is limited by θ, determined by the numerical aperture of the lens used

μm across) necessitates the use of focussing optics. Focussing optics limit the distance over which light is spatially coherent to λ/θ, where λ is the wavelength of incident light and the angle θ is small and defined in Fig. 4.1. This limits the number of nanostructures that can couple together in a 1D grating to $N = \frac{\lambda_R}{\theta a}$.

The first experimental realisations of extremely narrow diffraction-coupled plasmon resonances [10, 11] were achieved by using focussing optics with NA $= 0.1$, giving $\theta \approx 0.1$ rad in air and $N \sim 20$ for the main Rayleigh resonance in air. Kravets et al. [10] fabricated gold nanodot arrays on a glass substrate. Illuminating the arrays at oblique angles of incidence (62–68°) revealed extremely narrow resonances with FWHM <5 nm and $Q \sim 60$ in spectroscopic ellipsometry.

Meanwhile Auguié and Barnes [11] found narrow plasmon resonances in light transmitted through nanorod arrays fabricated on fused silica at normal incidence. In this case the resonances were only generated in a symmetric refractive index environment. An additional limiting factor on the quality of diffraction coupled plasmon resonances is any mismatch between the refractive index of the substrate and ambient medium, which can act to suppress lattice resonances [28].

4.3 Sample Design

Electron beam lithography (described in Sect. 2.3.2) was used to fabricate samples made from gold on a glass substrate. For all samples, substrates were coated with a 3 nm layer of chromium prior to lithography to prevent charging. In the first lithography step, a flat 100 μm × 300 μm region of 3 nm Cr (for adhesion) and 65 nm Au was fabricated on the substrate by electron beam evaporation. In a second lithographic step, a 300 μm array of 100 μm long, 450 nm wide gold stripes with a fixed height in the range from 60 to 90 nm were fabricated atop this flat gold region. An additional 3 nm of chromium was again deposited prior to the gold deposition step for adhesion. The periods, a, of the line arrays on each sample were measured with scanning electron microscopy (SEM) after fabrication and were around 1500 nm for the bulk of the samples (typical SEM micrograph shown in Fig. 4.2b). Control samples, consisting of just the lines of the second lithography step atop the plain glass/chromium

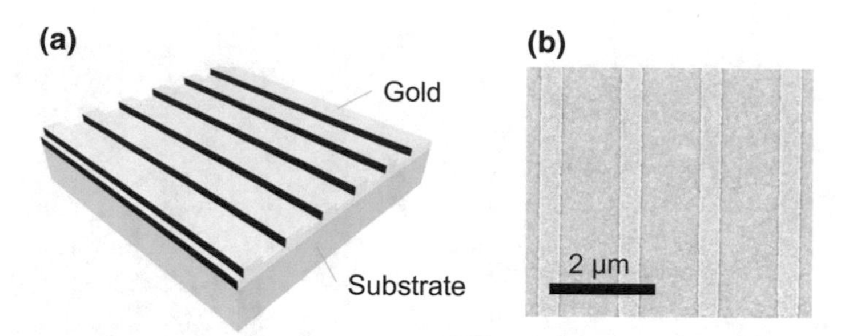

Fig. 4.2 Plasmonic nanostripe array. **a** 3D rendering of a structure that demonstrates extremely narrow plasmon resonances. **b** A SEM micrograph of the lines

substrate, were also fabricated and confirmed to not produce the extremely narrow resonances observed on the samples with the gold sublayer.

4.4 Results

We characterised our samples using spectroscopic ellipsometry as described in Sect. 2.2.5. The typical high quality plasmonic resonance based on diffraction coupled plasmons measured in our samples is shown in Fig. 4.3a. We estimate a quality factor $Q \approx 300$ at a wavelength of $\lambda = 1515$ nm at an angle of incidence of $\theta = 80°$, while the FWHM of the resonance is around 5 nm (our methodology for finding the FWHM and Q is described below). The sample has lines 420 nm wide and 70 nm high with a period $a = 1530$ nm and were fabricated on a 65 nm gold film. The resonance position corresponds very well to the expected position of the Rayleigh cutoff wavelength for air λ_R^{air} for $m = 2$. Peaks corresponding to the $m = 3, 4,$ and 5 resonance modes are also present in Fig. 4.3a. Figure 4.3b shows that the positions of the resonances closely match the Rayleigh cutoff wavelengths across a wide range of angles. It is important to stress that the control sets of stripes with the same sizes and periodicities fabricated on a bare glass substrate did not show narrow resonances at the same conditions; an example is shown in Fig. 4.3a. Therefore, the gold sublayer beneath the stripe array allowed us to suppress the negative effect of the substrate on collective resonances in an asymmetric refractive index environment.

The coupled resonances in plasmonic nanoarrays are highly asymmetric (the asymmetry comes in part from Fano interference and in part from disappearance of a diffractive order in analogy with total internal reflection), meaning that the estimated FWHM (and hence value of Q) depends strongly on which side of the resonance we use as a baseline. Taking the steeper right-hand side of the peak gives our best resonance a lower-bound $Q = 242$, whereas we estimate a much higher $Q = 362$ using the shallower left-hand side. All FWHMs and quality factors quoted hereafter are calculated using the average of these two estimation methods. This gives our best resonance quality observed in the telecom band as $Q \approx 300$.

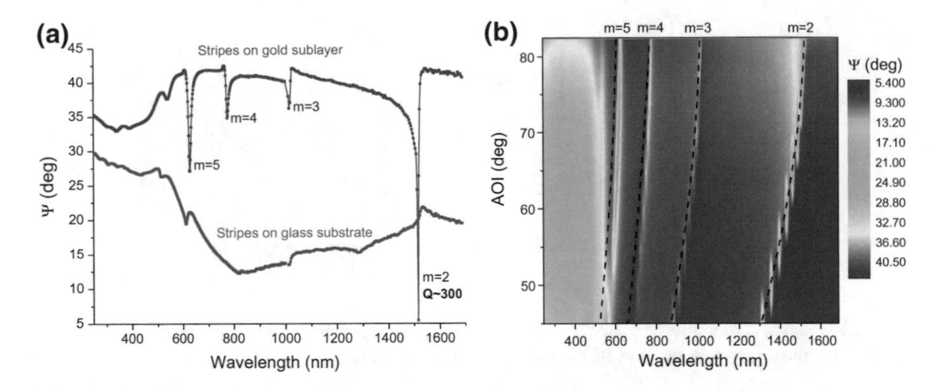

Fig. 4.3 Spectroscopic ellipsometry of diffraction coupled plasmon resonance measured in a nanostripe array with stripe width 420 nm wide, stripe period $a = 1530$ nm. **a** Typical spectrum at an angle of incidence (AOI) of 80°. The blue curve shows the collective resonances of stripes fabricated on a gold layer; the magenta curve gives the ellipsometric reflection for the control stripes made on bare glass substrate. **b** Spectroscopic ellipsometry of diffraction coupled plasmon resonances for AOIs $= 45 - 82.5°$. The dashed lines show the predicted Rayleigh cut-off wavelengths λ_R for $m = 2, 3, 4$ and 5

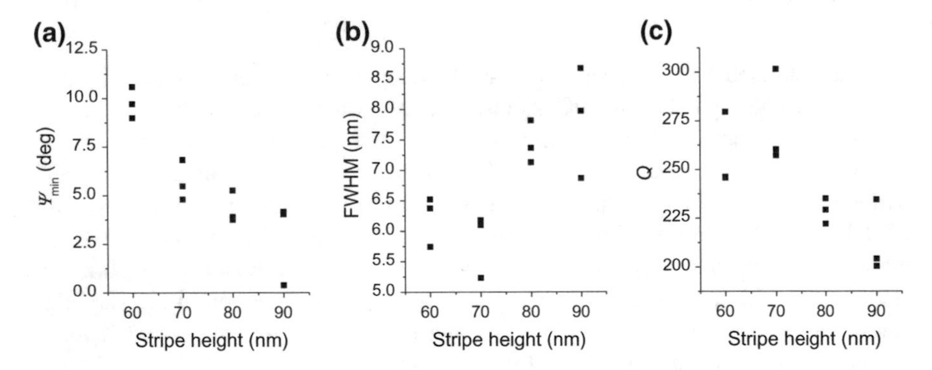

Fig. 4.4 Parameters of the collective resonances as a function of stripe height. **a** Reflection minimum at the resonance position Ψ_{min}, **b** FWHM and **c** the resonance quality Q for the Rayleigh mode $m = 2$ of diffraction coupled resonances

To study the effect of the stripe height on the quality of the collective resonances of the stripes fabricated on a thin gold layer, we fabricated stripe arrays with stripe heights of 60, 70, 80, and 90 nm. The thickness of the gold sublayer was kept constant at 65 nm. The minimum values of Ψ (Ψ_{min}), FWHM, and Q values for the $m = 2$ resonances of these arrays are plotted in Fig. 4.4. Figure 4.4a shows that Ψ_{min} decreases with increasing stripe height; in other words, increasing the stripe height makes the resonances deeper. However, Fig. 4.4b demonstrates that the FWHM increases with increasing stripe height (and so, as shown in Fig. 4.4c, Q correspondingly decreases). This implies that the stripe height of 70 nm was optimal for achieving the sharpest plasmonic resonances of the best quality, while the stripe height of 90 nm was

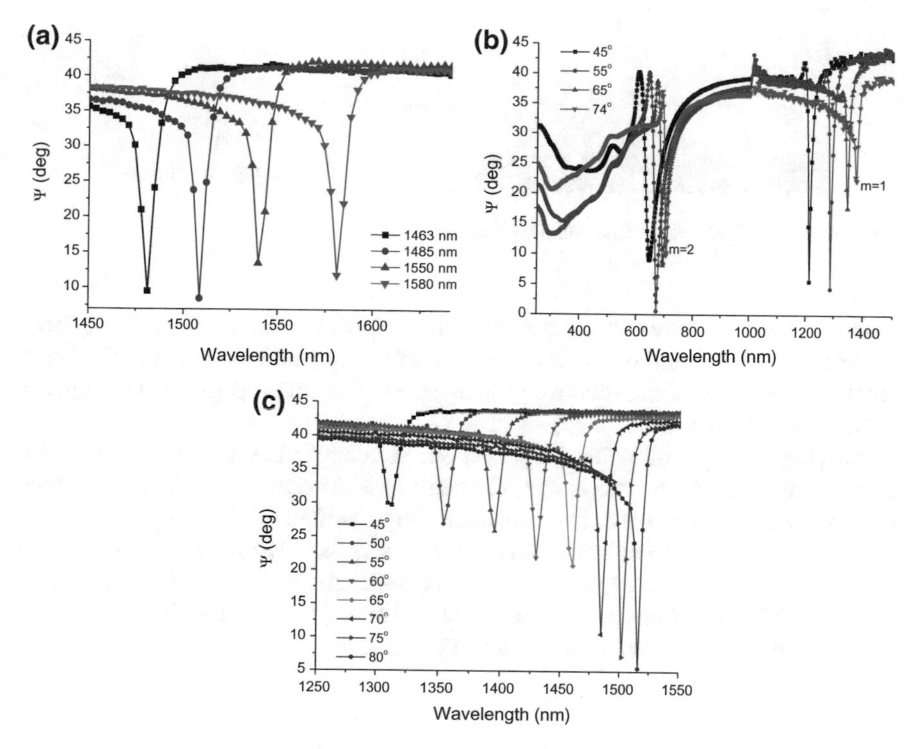

Fig. 4.5 Properties of collective resonances in nanostripe arrays. **a** Close up of the $m = 2$ resonance peaks of line arrays of various periods at angle of incidence of 80°. The lines were 450 nm wide in all cases and the quality factor $Q \sim 200$. **b** Narrow resonances at the near infrared corresponding to the $m = 1$ and $m = 2$ Rayleigh cut-off wavelengths measured on a sample with array period 700 nm, stripe width 200 nm. **c** Angle dependence of collective resonances. Spectra measured on a sample with period 1530 nm and stripe width 420 nm for angles of incidence from 45 to 80° in 5° steps

optimal for the deepest plasmonic resonances, which are required to realize high phase sensitivity [5].

The ability to tune the position of a surface plasmon resonance is critical for applications such as telecommunications and spectroscopy. Figure 4.5a[3] shows that it is possible to shift the narrow $m = 2$ resonance from 1480 nm up to 1580 nm while retaining the quality of the resonance by increasing the period of the array. Narrow resonances can be achieved using our device structure over an even wider range of array periods: Fig. 4.5b shows that narrow resonances with $Q \sim 130$ can be achieved using a period of $a = 700$ nm in the 1200 nm $< \lambda <$ 1300 nm region. These resonances correspond to the $m = 1$ Rayleigh cut-off wavelength. $m = 2$ modes also appear at

[3]Figure 4.5a and b are reproduced from Ben Thackray's preliminary studies of super-narrow plasmon resonances [1]. The $a = 1463$ nm and $a = 1485$ nm curves in Fig. 4.5a and all data in Fig. 4.5b are from stripe arrays fabricated and characterised by Ben Thackray (again with assistance from Gregory Auton for lithography) in the early stages of this study.

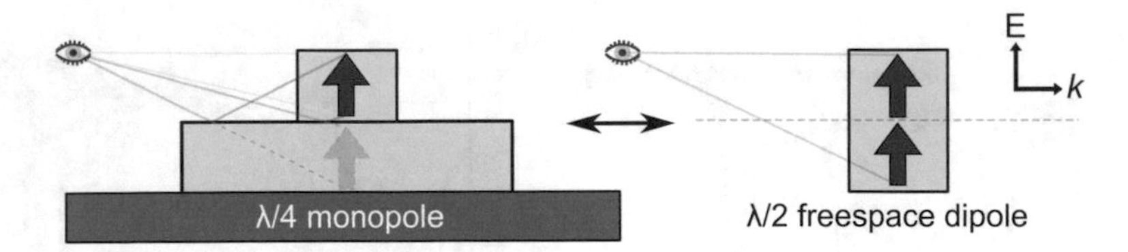

Fig. 4.6 Image dipoles due to the gold film underneath the stripes

around 650 nm; they are of a lower quality than resonances that have previously been achieved in this spectral region using arrays without a gold sublayer [6, 10]. These results suggest that we can effectively change the reflection/propagating modes in our devices by varying the periodicity of stripes.

Varying the angle of incident light on our structures allows for an extra degree of tunability. Figure 4.5c shows that, although Q is maximized for the resonance at an angle of incidence of 80°, the resonance can be shifted by changing the angle of incidence. The resonances remain deep over a surprisingly broad range of angles; only below 75° does the depth of the resonances dramatically decrease. Even then, at 45° the FWHM of the $m = 2$ resonance is 12 nm (compared with 7 nm at 80°), while the resonance moves from 1515 to 1311 nm.

4.5 Discussion

The unusually high quality of our plasmon resonances can be attributed to enhancements to the strong out-of-plane electron oscillations in our nanostripes. Out-of-plane electron oscillations have been extensively studied in a number of experimental and theoretical works [24, 28, 29]; in particular, they have been shown to allow coupling to the narrow subradiant lattice resonances [29].

The effect of the substrate on the out-of-plane oscillations in our nanostripe arrays is enhanced by the presence of the gold sublayer [30, 31]. The sublayer acts to effectively increase the height of the stripes. This effect can be understood by analogy to antenna theory, illustrated in Fig. 4.6. A monopole quarter-wave antenna over a perfect ground would radiate a field equivalent to that of a half-wave dipole antenna of twice the voltage. The condition for this effect to occur is that the gold sublayer is thicker than the skin depth of gold, which is ~5 nm for light at a wavelength of ~1.5 μm. Out-of-plane electron oscillations in the stripes give rise to image charges in the gold sublayer. The overall effect is that the scattered field from out-of-plane electron oscillations is similar to that of a structure of twice the height. We therefore expect the strongest, narrowest resonances at high angles of incidence (~80°) when the component of the electric field normal to the plane is maximised.

For s-polarised light the dipole and its image in the sublayer will be antiparallel and therefore cancel out, meaning this effect highly polarisation sensitive.

A more elaborate theory of the diffraction coupled resonances can be derived within the framework of the coupled dipole approximation or direct calculations with the help of finite difference time domain solvers as was done in many publications before [8, 9, 23, 29, 32–34].

Another factor responsible for the sharpness of our resonances is the creation of a "virtual" index-matched environment. The presence of the sublayer means that light responsible for diffraction coupling does not penetrate into substrate. This negates the damping effect the substrate refractive index mismatch usually has on the diffraction coupled resonances [28].

4.6 Conclusion

Diffraction coupled plasmon resonances of exceptionally narrow spectral width and high quality factors were achieved by fabricating arrays of gold nanostripes on a gold sublayer. The effect is explained by analogy with well-known results from the theory of antennae and attributed to image charges in the gold sublayer, which create a radiation field equivalent to that of a resonant nanostructure of twice the height. This, coupled with excitation at a steep angle of incidence, creates strong out-of-plane electron oscillations. The gold sublayer also acts to mimic an index-matched environment and remove the damping effect of an asymmetric environment. The measured quality factors of up to 300 are among the highest reported from arrays of diffraction coupled nanostructures at wavelengths around 1.5 μm.

References

1. B.D. Thackray, Coupling of localised plasmon resonances. Ph.D. thesis, The University of Manchester, Manchester, UK, 2014
2. B.D. Thackray, P.A. Thomas, G.H. Auton, F.J. Rodriguez, O.P. Marshall, V.G. Kravets, A.N. Grigorenko, Super-narrow, extremely high quality collective plasmon resonances at telecom wavelengths and their application in a hybrid graphene-plasmonic modulator. Nano Lett. **15**(5), 3519–3523 (2015)
3. J.B. Pendry, Negative refraction makes a perfect lens. Phys. Rev. Lett. **85**(18), 3966 (2000)
4. J.N. Anker, W.P. Hall, O. Lyandres, N.C. Shah, J. Zhao, R.P. Van Duyne, Biosensing with plasmonic nanosensors. Nat. Mater. **7**(6), 442–453 (2008)
5. V.G. Kravets, F. Schedin, R. Jalil, L. Britnell, R.V. Gorbachev, D. Ansell, B. Thackray, K.S. Novoselov, A.K. Geim, A.V. Kabashin, A.N. Grigorenko, Singular phase nano-optics in plasmonic metamaterials for label-free single-molecule detection. Nat. Mater. **12**(4), 304–309 (2013)
6. V.G. Kravets, R. Jalil, Y.-J. Kim, D. Ansell, D.E. Aznakayeva, B. Thackray, L. Britnell, B.D. Belle, F. Withers, I.P. Radko, Z. Han, S.I. Bozhevolnyi, K.S. Novoselov, A.K. Geim, A.N. Grigorenko, Graphene-protected copper and silver plasmonics. Sci. Rep. **4** (2014)
7. J. Kim, H. Son, D.J. Cho, B. Geng, W. Regan, S. Shi, K. Kim, A. Zettl, Y.R. Shen, F. Wang, Electrical control of optical plasmon resonance with graphene. Nano Lett. **12**(11), 5598–5602 (2012)

8. S. Zou, N. Janel, G.C. Schatz, Silver nanoparticle array structures that produce remarkably narrow plasmon lineshapes. J. Chem. Phys. **120**(23), 10871–10875 (2004)
9. V.A. Markel, Divergence of dipole sums and the nature of non-lorentzian exponentially narrow resonances in one-dimensional periodic arrays of nanospheres. J. Phys. B Atomic Mol. Optical Phys. **38**(7), L115 (2005)
10. V.G. Kravets, F. Schedin, A.N. Grigorenko, Extremely narrow plasmon resonances based on diffraction coupling of localized plasmons in arrays of metallic nanoparticles. Phys. Rev. Lett. **101**(8), 087403 (2008)
11. B. Auguié, W.L. Barnes, Collective resonances in gold nanoparticle arrays. Phys. Rev. Lett. **101**(14), 143902 (2008)
12. L. Rayleigh, On the dynamical theory of gratings. Proc. R. Soc. Lond. Ser. A Contain. Papers Math. Phys. Character **79**(532), 399–416 (1907)
13. S.-Q. Li, W. Zhou, D. Bruce Buchholz, J.B. Ketterson, L.E. Ocola, K. Sakoda, R.P.H. Chang, Ultra-sharp plasmonic resonances from monopole optical nanoantenna phased arrays. Appl. Phys. Lett. **104**(23), 231101 (2014)
14. F. Wang, Y.R. Shen, General properties of local plasmons in metal nanostructures. Phys. Rev. Lett. **97**(20), 206806 (2006)
15. V.G. Kravets, F. Schedin, A.N. Grigorenko, Fine structure constant and quantized optical transparency of plasmonic nanoarrays. Nat. Commun. **3**, 640 (2012)
16. N.J. Halas, S. Lal, W.-S. Chang, S. Link, P. Nordlander, Plasmons in strongly coupled metallic nanostructures. Chem. Rev. **111**(6), 3913–3961 (2011)
17. A.L. Koh, K. Bao, I. Khan, W.E. Smith, G. Kothleitner, P. Nordlander, S.A. Maier, D.W. McComb, Electron energy-loss spectroscopy (eels) of surface plasmons in single silver nanoparticles and dimers: influence of beam damage and mapping of dark modes. ACS Nano **3**(10), 3015–3022 (2009)
18. V.A. Markel, Coupled-dipole approach to scattering of light from a one-dimensional periodic dipole structure. J. Mod. Optics **40**(11), 2281–2291 (1993)
19. S. Zou, G.C. Schatz, Theoretical studies of plasmon resonances in one-dimensional nanoparticle chains: narrow lineshapes with tunable widths. Nanotechnology **17**(11), 2813 (2006)
20. H. DeVoe, Optical properties of molecular aggregates. I. Classical model of electronic absorption and refraction. J. Chem. Phys. **41**(2), 393–400 (1964)
21. H. DeVoe, Optical properties of molecular aggregates. II. Classical theory of the refraction, absorption, and optical activity of solutions and crystals. J. Chem. Phys. **43**(9), 3199–3208 (1965)
22. E.M. Purcell, C.R. Pennypacker, Scattering and absorption of light by nonspherical dielectric grains. Astrophys. J. **186**, 705–714 (1973)
23. F.J. García De Abajo, Colloquium: light scattering by particle and hole arrays. Rev. Mod. Phys. **79**(4), 1267 (2007)
24. B.D. Thackray, V.G. Kravets, F. Schedin, G. Auton, P.A. Thomas, A.N. Grigorenko, Narrow collective plasmon resonances in nanostructure arrays observed at normal light incidence for simplified sensing in asymmetric air and water environments. ACS Photon. **1**(11), 1116–1126 (2014)
25. C.L. Haynes, A.D. McFarland, L. Zhao, R.P. Van Duyne, G.C. Schatz, L. Gunnarsson, J. Prikulis, B. Kasemo, M. Käll, Nanoparticle optics: the importance of radiative dipole coupling in two-dimensional nanoparticle arrays. J. Phys. Chem. B **107**(30), 7337–7342 (2003)
26. E.M. Hicks, S. Zou, G.C. Schatz, K.G. Spears, R.P. Van Duyne, L. Gunnarsson, T. Rindzevicius, B. Kasemo, M. Käll, Controlling plasmon line shapes through diffractive coupling in linear arrays of cylindrical nanoparticles fabricated by electron beam lithography. Nano Lett. **5**(6), 1065–1070 (2005)
27. J. Sung, E.M. Hicks, R.P. Van Duyne, K.G. Spears, Nanoparticle spectroscopy: plasmon coupling in finite-sized two-dimensional arrays of cylindrical silver nanoparticles. J. Phys. Chem. C **112**(11), 4091–4096 (2008)
28. B. Auguié, X.M. Bendana, W.L. Barnes, F.J. García de Abajo, Diffractive arrays of gold nanoparticles near an interface: critical role of the substrate. Phys. Rev. B **82**(15), 155447 (2010)

29. W. Zhou, T.W. Odom, Tunable subradiant lattice plasmons by out-of-plane dipolar interactions. Nat. Nanotechnol. **6**(7), 423–427 (2011)
30. T. Yamaguchi, S. Yoshida, A. Kinbara, Optical effect of the substrate on the anomalous absorption of aggregated silver films. Thin Solid Films **21**(1), 173–187 (1974)
31. V.G. Kravets, F. Schedin, S. Taylor, D. Viita, A.N. Grigorenko, Plasmonic resonances in optomagnetic metamaterials based on double dot arrays. Optics Express **18**(10), 9780–9790 (2010)
32. S. Mubeen, S. Zhang, N. Kim, S. Lee, S. Krämer, H. Xu, M. Moskovits, Plasmonic properties of gold nanoparticles separated from a gold mirror by an ultrathin oxide. Nano Lett. **12**(4), 2088–2094 (2012)
33. M.G. Moharam, T.K. Gaylord, Rigorous coupled-wave analysis of metallic surface-relief gratings. JOSA A **3**(11), 1780–1787 (1986)
34. N. Papanikolaou, Optical properties of metallic nanoparticle arrays on a thin metallic film. Phys. Rev. B **75**(23), 235426 (2007)

Chapter 5
Nanomechanical Electro-Optical Modulator Based on Atomic Heterostructures

Two-dimensional atomic heterostructures combined with metallic nanostructures allow one to realize strong light–matter interactions. Metallic nanostructures possess plasmonic resonances that can be modulated by graphene gating. In particular, spectrally narrow plasmon resonances potentially allow for very high graphene-enabled modulation depth. However, the modulation depths achieved with this approach have so far been low and the modulation wavelength range limited. Here we demonstrate a device in which a graphene/hexagonal boron nitride heterostructure is suspended over a gold nanostripe array. A gate voltage across these devices alters the location of the two-dimensional crystals, creating strong optical modulation of its reflection spectra at multiple wavelengths: in ultraviolet Fabry-Perot resonances, in visible and near-infrared diffraction-coupled plasmonic resonances and in the mid-infrared range of hexagonal boron nitride's upper Reststrahlen band. Devices can be extremely subwavelength in thickness and exhibit compact and truly broadband modulation of optical signals using heterostructures of two-dimensional materials.

The devices fabricated and studied in this chapter consist of atomic heterostructures transferred on top of the gold nanostripe arrays described in Chap. 4. I designed the general device structure; the fabrication and transfer of the heterostructures was performed by Francisco Rodriguez and Gregory Auton. I characterised the devices using ellipsometry; additional characterisation using Raman spectroscopy, FTIR, atomic force microscopy (AFM), optical reflection and frequency measurements was done in collaboration with Vasyl Kravets, Owen Marshall, Yang Su, Dmytro Kundys and Francisco Rodriguez, respectively. I performed calculations to estimate the Maxwell stresses and the upper limit of the modulation frequency. Owen Marshall performed modelling of the optical response of the device. I coauthored the manuscript with Owen Marshall and Alexander Grigorenko.

P. A. Thomas, *Narrow Plasmon Resonances in Hybrid Systems*,
Springer Theses, https://doi.org/10.1007/978-3-319-97526-9_5

5.1 Introduction

Plasmonics and the nano-optics of two-dimensional (2D) materials are two of the most rapidly-advancing areas in photonics [1–3]. The advent of high-resolution nanofabrication techniques has allowed for the development of plasmonic nanostructures with a broad range of optical responses such as subwavelength confinement of light [4] and ultra-narrow diffraction-coupled resonances [5–7], which could lead to applications in waveguiding [8] and biosensing [9], respectively. Meanwhile, the extraordinary properties of 2D materials have generated great interest in the nano-optics community [10]. Graphene has demonstrated remarkable optical properties which include visual transparency defined by the fine structure constant [11] and gate-tuneable intrinsic plasmons [3, 12, 13]. More recently, graphene has been combined with other 2D materials such as hexagonal boron nitride (hBN) to form heterostructures [14], which have already been used to develop proof-of-concept devices such as light-emitting diodes [15].

The combination of plasmonic nanostructures with 2D heterostructures shows strong promise [3]. Placing plasmonic nanostructures in 2D heterostructure stacks allows for much stronger interactions between light and 2D materials [16]. This approach has been used to produce significant advances in the field of photovoltaics [17], Raman spectroscopy [18] and sensing [9]. Arguably the most exciting developments could come in optoelectronics [2, 3, 8]—the optical properties of graphene can easily be tuned by applying a gate voltage [8]. Combining graphene with metallic nanostructures capable of supporting ultra-narrow plasmon resonances could potentially lead to the development of compact optical modulators [3]. However, attempts to experimentally realise such a device have met with limited success [19–22]—modulation depths are typically small (or slow for graphene supercapacitors) and influence over higher-energy spectral features requires exceptionally high doping of graphene.

Here we use electromechanical modulation to achieve high modulation depths in a larger frequency range. Indeed, relatively little attention has been paid to the potential for nanoelectromechanical optical effects in either plasmonic or 2D heterostructure systems [23, 24]. We describe a graphene/hBN/ gold nanostripe heterostructure with an air gap between the hBN and plasmonic nanostripes. Application of a gate voltage between graphene and the gold nanostripe sublayer leads to a reduction in the air gap height. This in turn changes the effective electric field acting on graphene and hBN, modifying spectral reflection features from mid-ultraviolet to mid-infrared wavelengths. The demonstration of strong modulation (of up to 30%) in a 2D material-based device, especially over such a broad range of spectral features, is unprecedented. It is important to stress that the volume in which the modulation is achieved is extremely small. For example, in the mid-infrared region (at a wavelength $\lambda \sim 7\ \mu$m) the optical interrogation volume is $\sim\lambda^3/10$, the smallest reported so far—approximately three orders of magnitude smaller than reported in [24]. Our results open up new possibilities for strong light modulation in optoelectronic devices.

5.2 Sample Design

The design of our graphene/hBN/nanoarray modulator is shown in Fig. 5.1a. The gold plasmonic nanostripe array is separated from the graphene/hBN by an air gap (d). Using the graphene as a broadband, transparent and extremely tough electrical contact, a gate voltage can be applied across the structure. In this work the nanostripe array had a periodicity $a = 1570$ nm, stripe width $w = 550$ nm, stripe height $h_2 = 80$ nm and gold sublayer of thickness $h_1 = 65$ nm (Fig. 5.1b). Uniform arrays of gold nanostripes (total area = 300 μm $\times$ 100 μm) were fabricated (with a gold sublayer) on glass substrates using standard electron beam lithography and electron beam evaporation techniques (see Sect. 2.3.2). First, the glass substrate was covered with a thin film of Cr (3 nm) to prevent charging during lithography, followed a flat region of Cr (3 nm, for adhesion) and Au (65 nm). As well as forming the metallic sublayer, this region acted as the back contact for electrical measurements. Next, the array of gold nanostripes (length 100 μm) was fabricated on top of the gold sublayer, once again using a Cr (3 nm) adhesion layer. Finally, the exposed areas of the initial Cr film were removed via wet chemical etching. Nanostripe dimensions were confirmed using scanning electron microscopy. A representative SEM micrograph is shown in Fig. 5.1c. Such plasmonic nanoarrays can be tuned to give narrow, diffraction-coupled resonances which arise when the wavelengths of diffracted light modes, running along the air-substrate boundary (known as Rayleigh cut-off wavelengths), are recaptured as electron oscillations in the plasmonic nanostructures. As described in Chap. 4 these resonances can be further narrowed by adding a metallic sublayer. Our nanostructure was designed to produce a narrow plasmon resonance around the telecom wavelength of $\lambda \sim 1.5$ μm, although higher-order diffraction-coupled modes exist throughout the near-infrared and visible spectrum.

An hBN flake of ($\sim$110 nm thick) and single-layered graphene were then mechanically exfoliated and transferred on to the plasmonic nanostructure using the wet transfer technique (described in Sect. 3.4.1). hBN is an ideal dielectric for graphene devices because it is an atomically flat crystal with a similar lattice constant to graphene [25]. Adhesion between the hBN and rough nanoarray surface was non-uniform over the flake area, leading to the existence of the required air gap in regions of each device. The nanostripe array period is significantly larger than width of the individual nanostripes, resulting in a small contact area between the hBN and nanoarray surface. The gaps in physical contact have a natural tendency to propagate, leading to a high probability of the formation of air gaps (bubbles) during heterostructure transfer.

Figure 5.2a provides an optical image of the graphene/hBN stack on gold nanostripe array studied in our experiments. We can see that the area of the suspended stack is approximately rectangular, although the clamped edges of the stack have a more complicated form. To assess the height profile of the suspended stack we performed AFM measurements, as shown in Fig. 5.2b and c. Figure 5.2b gives the AFM image of a sub-section of the device area, with Fig. 5.2c showing the height line-scan along the dashed line in Fig. 5.2b. Along the length of this line-scan we can clearly resolve

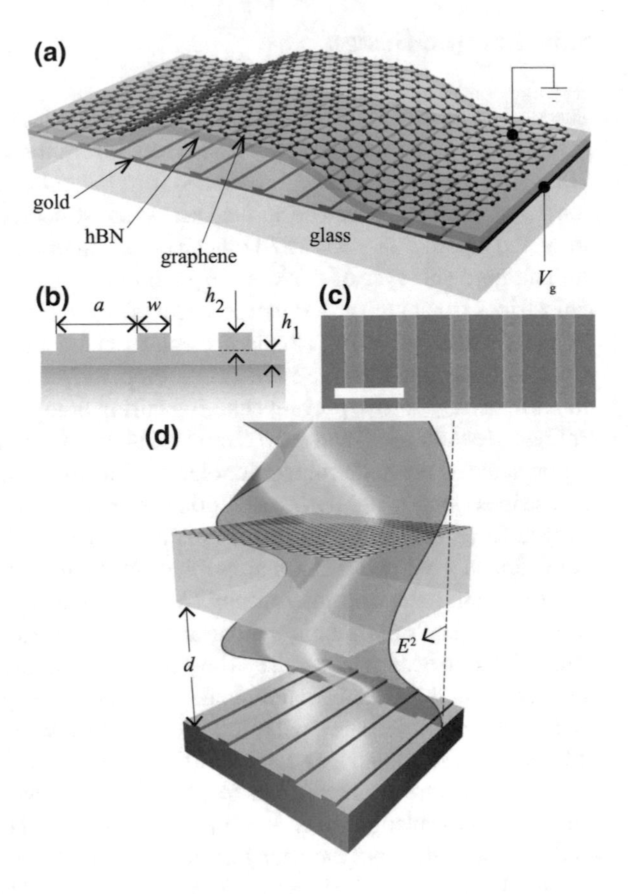

Fig. 5.1 Nanomechanical electro-optical modulator structure. **a** Schematic of our device with air gap height d. **b** Geometric design parameters for our gold nanostripe array. **c** Representative SEM micrograph of the nanostripe array (scale bar 2 μm). **d** The working principle of the device. The coloured wave represents an unperturbed standing wave for different wavelengths observed under reflection from the nanostructured mirror

the individual gold stripes in the uncovered region, as well as the area of hBN in contact with the gold stripes and the suspended region. The maximum measured air gap was around 350 nm, which is in good agreement with estimates we obtained from optical spectra fitting (described in Sect. 5.3.1).

The purple circles in Fig. 5.2c show the theoretical fitting of the hBN profile using a simple model of a clamped thin rectangular membrane [26]. Since we have a good agreement between the calculated and the measured profile, we apply this simple model to quantitatively evaluate the change of the membrane height due to applied voltage. The maximal height of the clamped thin rectangular membrane with a smaller side (length b) and a larger side (length a) can be evaluated as

$$w = C \cdot \left(1 - \nu^2\right) \frac{(p_0 - p_a) \cdot b^4}{E \cdot d^3}, \tag{5.1}$$

where $C = 0.32/(1 + (b/a)^4)$, ν is the Poisson ratio, p_0 is the initial pressure inside the bubble, p_a is the atmospheric pressure, E is the Young modulus, and d is the membrane thickness. In our case the length of the smaller side is around $b \approx 50$ μm,

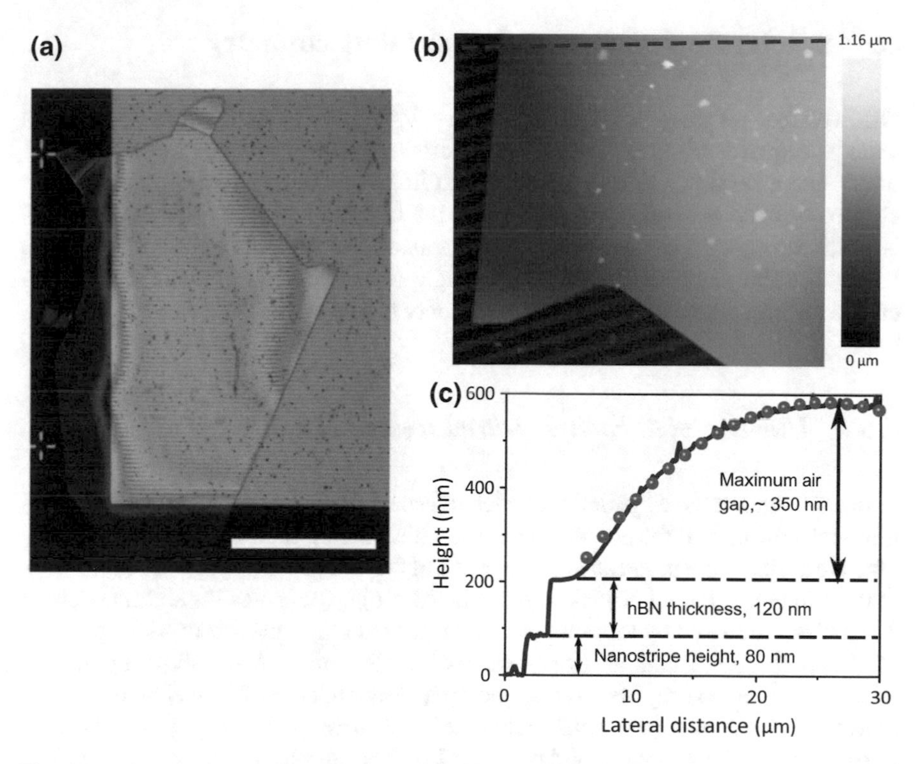

Fig. 5.2 **a** Optical image of graphene/hBN stack on gold nanostripe array. Scale bar 50 μm. Interference fringes indicative of an air gap are clearly visible. The presence of single layered graphene over the entire hBN area was confirmed with Raman spectroscopy. **b** AFM image of edge region of the graphene/hBN stack on gold nanostripe array. **c** Profile along dashed line in (**b**). The purple circles show the theoretical fit for a clamped thin membrane

$b/a \approx 0.5$, $\nu \approx 0.3$ and $w \approx 350$ nm, which yields the initial pressure inside the bubble $p_0 \approx 1.2$ atm. When the gate voltage is applied to graphene, the height of the membrane would decrease due to the increased top pressure by the electrostatic pressure p_e (evaluated in Sect. 5.5.1) and counteracted by the increased pressure inside the bubble $\tilde{p}_0$ (due to the decreased bubble volume). The new height $\tilde{w}$ can be found from an implicit relation

$$\tilde{w} = C \cdot \left(1 - \nu^2\right) \frac{(\tilde{p}_0 - p_e - p_a) \cdot b^4}{E \cdot d^3}, \tag{5.2}$$

where $\tilde{p}_0 = p_0(w - d)/(\tilde{w} - d)$. Solving these two coupled equations we obtain the change in the height of the bubble $w - \tilde{w} \approx 200$ nm at an electrostatic pressure of 20 atm, which is comparable with the estimates we obtained from optical spectra fitting. It is necessary to stress that the estimate is quite rough due to the simplicity of the model and the possible presence of hydrocarbons and water inside the bubble.

5.3 Spectroscopic Ellipsometry and Reflectometry

Gate voltages were applied using a Keithley 2400 Sourcemeter. All measurements were performed under standard room temperature and pressure conditions. Ellipsometric and reflection spectra were measured in a Woollam M-2000F focused beam ellipsometer (Numerical Aperture 0.1). At the highest angles of incidence (~80°), with the strongest localised plasmon resonances, the measurement spot size was larger than the graphene/hBN area; the angle of incidence was reduced to 70° to ensure the measurement spot was incident only on the active device area.

5.3.1 *Ultraviolet to Near-Infrared Response*

Figure 5.3a shows the ellipsometric reflection spectrum (ψ) of a device from the mid-ultraviolet through to the near-infrared when illuminated at an incident angle of $\theta = 70°$. We attribute the broad absorption peaks in the wavelength range 280 nm $< \lambda <$ 590 nm to Fabry-Perot (FP) interference in the air gap, whereas the sharp feature at $\lambda = 275$ nm is caused by the complex, multi-peaked ultraviolet absorption spectrum of hBN [27]. The remaining strong features from 590 nm $< \lambda <$ 1600 nm primarily arise from the nanoarray and correspond to the Rayleigh cut-off wavelengths for air, determined by $\lambda = \frac{a}{m}(n + \sin\theta)$, where a is the array periodicity, m is a positive integer, and n the refractive index of air [5]. The absorption peaks at $\lambda \approx 620$, 780, 1030 and 1520 nm can be associated with the $m = 5$, 4, 3 and 2 diffraction-coupled modes, respectively. The $m = 4, 5$ features each consist of two peaks due to a mismatch between the inverse polarisability and retarded dipole sum of individual nanoparticles in the plasmonic nanoarray in this spectral region (see Sect. 4.2.1), caused by the presence of the hBN.

Figure 5.3b plots ψ as a function of both wavelength and gate voltage V_g, showing how the various features from the ultraviolet to near-infrared respond to electrical biasing of the device. As V_g is raised from 0 V to ±150 V we see dramatic changes in the FP resonances, along with a redshift of the Rayleigh resonance wavelengths. These changes in reflection happen due to motion of the hBN/graphene heterostructure—the applied gate voltage creates an electrostatic force within the device, acting to reduce d. The Maxwell stresses caused by induced electrical charges can be of the order of 10 atmospheres (see Sect. 5.5.1). At this length scale Casimir interactions between the graphene and gold are negligible [28]. It is worth noting that, in general, the opto-electro-mechanical response of the structure is symmetric with respect to the sign of V_g and there is a threshold of ±50 V before which application of V_g produces no change in optical reflection (no motion of the heterostructure). One exception to the above mechanism occurs near $\lambda = 1.6$ μm for large negative V_g. In this region the absorption changes due to electrical gating of graphene, moving its charge neutrality point and the spectral onset of optical Pauli blocking [3]. This results in an increase in the measured value of ψ at the highest wavelengths for

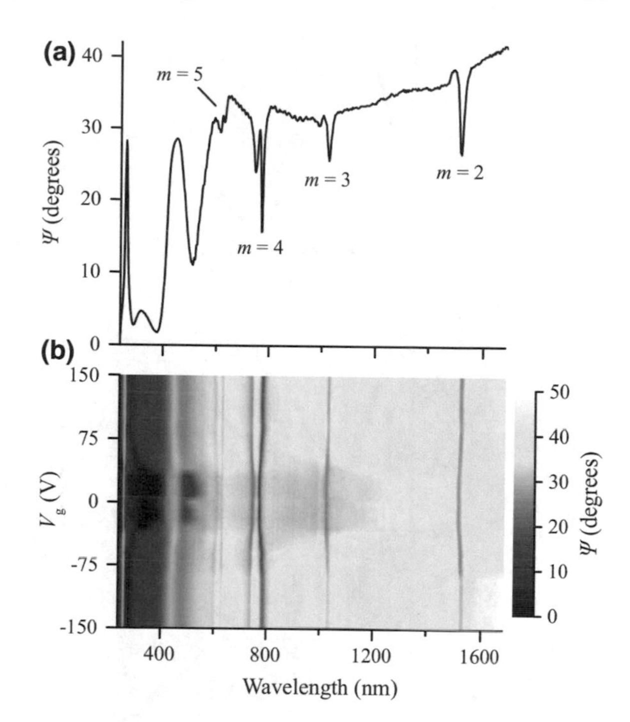

Fig. 5.3 Device characterisation using spectroscopic ellipsometry. **a** Ellipsometric reflection spectrum ψ of our graphene/boron nitride/plasmonic heterostructure ($V_g = 0$ V, $\theta = 70°$). **b** Colour map of ψ as a function of wavelength and V_g

the largest V_g. One can distinguish the effect of hBN motion and that of the Pauli blocking by the symmetry of the response (the effect of hBN motion is symmetric with respect to the sign of the applied voltage while the Pauli blocking effect is not symmetric due to the initial doping of graphene) and by the wavelength range at which these effects are observed, as described below.

The response to V_g in the ultraviolet region measured in s-polarised reflection (R_s) is shown in Fig. 5.4a. Note that the hBN absorption feature at $\lambda = 275$ nm is relatively insensitive to V_g (and hence d) as the wavelength is approximately equal to the optical path length within the hBN. However, as V_g is increased (and d decreased) the adjacent FP absorption features are dramatically altered. This cavity modulation effect occurs over the entire measured ultraviolet range ($\lambda > 250$ nm) and gives a peak modulation depth of $\sim$10% at $\lambda \sim 310$ nm. This compares favourably with prior attempts at ultraviolet modulation, which have usually relied on wide band gaps [29] or excitonic effects in semiconductor devices [30].

Moving to the visible and near-infrared ranges (for the same device) we have observed strong modulation of the diffraction-coupled resonances in the plasmonic nanoarray produced by graphene gating. Figure 5.4b shows the changes in p-polarized reflection spectra for the $\lambda = 780$ nm resonance, whereas Fig. 5.4c and d show ψ around 1030 nm and 1520 nm respectively. In all three cases the reflection minima are shifted by up to 10 nm for $|V_g| = 150$ V—as the hBN moves closer to the nanoarray

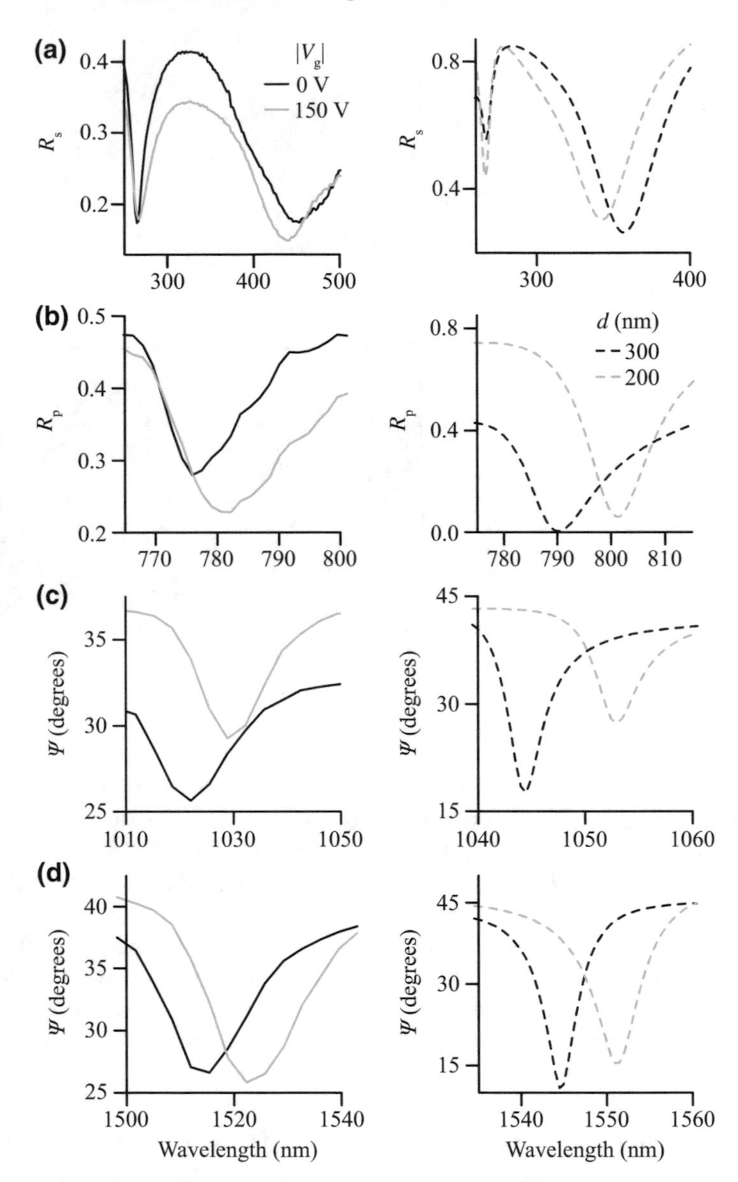

Fig. 5.4 Electromechanical modulation from ultraviolet to near-infrared regions. Measured reflection spectra (left column) of the graphene/hBN/nanoarray for **a** s-polarised light (R_s) in the ultraviolet/blue range and **b** p-polarised light (R_p) in the visible. **c** and **d** Ellipsometric reflection in the near-infrared. All spectra were measured with $\theta = 70°$. Modelling results (right column) reveal that the electrically induced reflection modulation is explained by a changing air gap. The sharp reflection minima in a (centred at $\lambda = 275$ nm) stems from the complex ultraviolet absorption spectrum of hBN. The features in b, c and d correspond to the fourth-, third- and second-order Rayleigh cut-off wavelengths of the nanoarray respectively

the local fields increase and the refractive index of the ambient medium (ordinarily $n = 1$ for air) is effectively changed. This in itself produces a strong modulation effect, simply because the plasmonic resonances are so narrow. Even though the minimum ψ value in the vicinity of $\lambda = 1520$ nm (Fig. 5.4d) remains approximately constant (~26–27°), its shifting wavelength with increasing $|V_g|$ leads to a modulation depth of 20% at $\lambda = 1536$ nm. This is significantly higher than the near-infrared absolute modulation depths previously reported in graphene- and plasmonic-based devices [22, 31–33]: for example, Li et al. [33] reported an absolute modulation depth of ~9% at telecom wavelengths using a graphene-clad microfibre.

5.3.1.1 Modelling Using Rigorous Coupled Wave Analysis

Reflection spectra obtained experimentally are described well by Rigorous Coupled Wave Analysis (RCWA) modelling, in which the graphene/hBN heterostructure is displaced vertically, reducing d from 300 to 200 nm.

Simulations were performed using the Reticolo package in Matlab [34], which relies upon a RCWA method [35]. The optical properties of Au were calculated within this package using a Lorentz-Drude model and the dispersion relation of hBN was calculated using a Lorentz oscillator model [36]:

$$\epsilon(\omega) = \epsilon_\infty \left(1 + \frac{\omega_{\mathrm{LO}}^2 - \omega_{\mathrm{TO}}^2}{\omega_{\mathrm{TO}}^2 - \omega^2 - i\gamma\omega}\right), \tag{5.3}$$

where ω_{LO} and ω_{TO} are the longitudinal optical and transverse optical phonon frequencies respectively, ϵ_∞ is the high frequency permittivity and γ is the damping constant associated with the plasma oscillation. An additional spectral absorption feature was introduced to the modelled hBN in order to reproduce the experimentally observed absorption peak in the near-ultraviolet region. Graphene optical conductivity was taken in the random-phase approximation [12].

The right-hand panels in Fig. 5.4a–d show the results of this modelling in the ultraviolet, visible and near-infrared regions of interest. Both the functional form and behaviour of the blue-shifting FP and red-shifting Rayleigh resonances are reproduced by the model. The differences between the experimental and modelled spectra can be attributed to impurities in, and roughness of, our gold nanostripe arrays. In addition, the height of the suspended area is not constant over the beam spot which leads to resonances smearing. Figure 5.5 shows further modelling of even smaller air gaps, revealing that as d approaches zero the Rayleigh resonance wavelengths become increasingly sensitive to changes in d. With this knowledge future devices might be deliberately engineered with smaller initial air gaps, potentially greatly enhancing the achievable modulation depths.

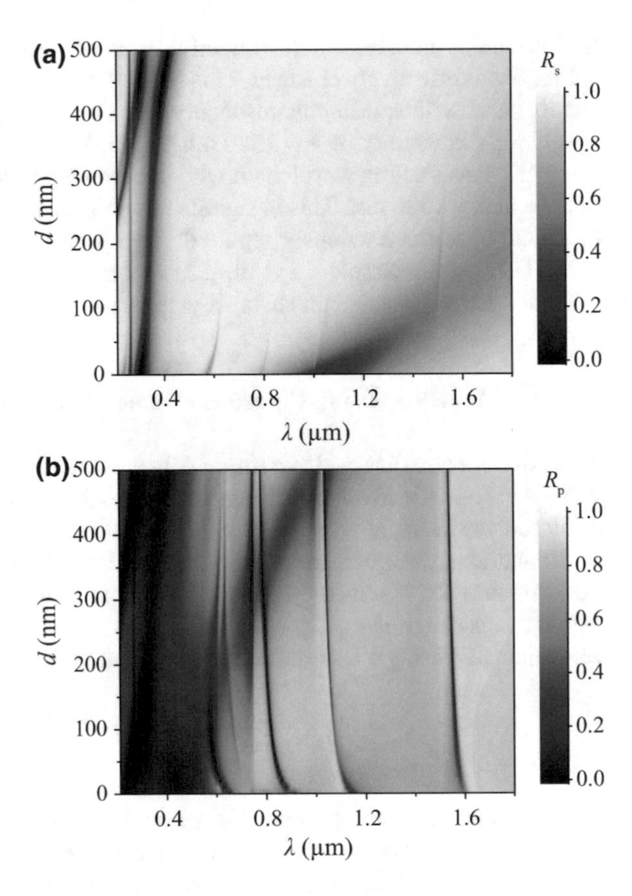

Fig. 5.5 Modelled spectral reflection in the λ-d plane. Reflection for **a** s-polarised and **b** p-polarized incident light. Angle of incidence $\theta = 70°$. For $d < 50$ nm the system becomes extremely sensitive to changes in d, leading to dramatically increased modulation depths

5.3.1.2 Effect of Optical Pauli Blocking

In our modelling we have used values for the Fermi energy calculated with the help of a simple capacitance model. In order to check these calculations we have measured gated optical spectra of our devices at low gate voltages such that the onset of motion in the graphene/hBN stack was not yet reached due to mechanical hysteresis. As a consequence, the optical reflection of the device only changes due to the effect of optical Pauli blocking.

Figure 5.6 shows the spectral dependence of relative reflection of our device in the wavelength range 2-7 μm (the relative reflection was measured as $R(V_g)/R(0)$). The black and red curves show the relative spectra at $V_g = 20$ V and $V_g = -20$ V. Since the initial doping of our exfoliated graphene was high, these spectral curves are close to unity, which is characteristic of the forbidden interband transition (and negligible intraband contribution). However, at $V_g = -50$ V we see change in the reflection centred around twice the Fermi energy position (at a wavelength of 3.8 μm) which is in a good agreement with a simple capacitance model for the Pauli blocking (provided any initial doping is taken into account).

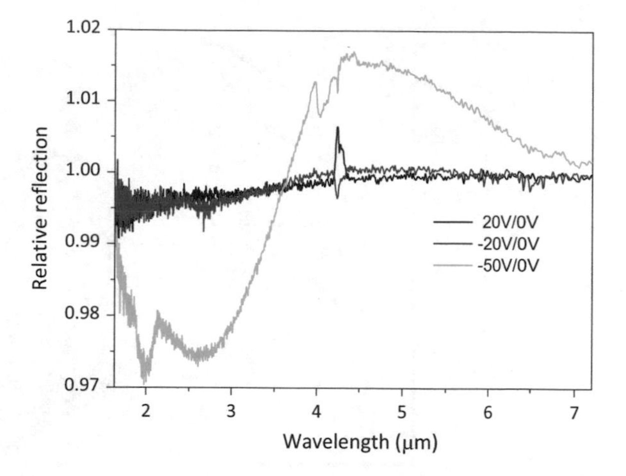

Fig. 5.6 Relative reflection showing effect of Pauli blocking in graphene. Relative reflection of our device in the wavelength range 2–7 μm for $V_g \pm 20$ V, -50 V

5.3.2 *Mid-infrared Response*

FTIR spectroscopy was performed using a Bruker Vertex 80 system with a Hyperion 3000 microscope. A variety of sources and detectors, combined with aluminium coated reflective optics enable this system to be used from visible to THz wavelengths. A reflecting objective provides a range of incident angles ($\theta = 12$–$24°$) and the entire beam path purged with dry, CO_2 scrubbed air to minimise strong atmospheric infrared absorption bands.

Figure 5.7a shows the strong modulation feature observed in the mid-infrared wavelength range produced by graphene gating. At such long wavelengths incident light is not influenced by the grating structure, instead effectively experiencing a planar gold surface. On the long wavelength side of its Reststrahlen band, just beyond the transverse optic phonon wavelength of ~7.35 μm, hBN displays strong light absorption and spectral dispersion. In this region we observe a broad reflection minimum, with reflection values falling to 33% at $\lambda \sim 7.6$ μm (Fig. 5.7a). The electromechanical reduction of d induces a narrowing of this feature and a blue shift of the reflection minimum by over 100 nm. The reflection minimum also falls to ~15%. As a result, large absolute reflection modulation depths are possible for a given wavelength—up to 30% at $\lambda = 7.5$ μm. This represents a dramatic improvement on existing mid-infrared graphene-plasmonic modulator results [21, 37, 38]: for example, Yao et al. [38] have previously reported an absolute reflection modulation depth of ~20% around 7.6 μm. Comparing the device dimensions (total height ~450–550 nm) with the wavelength of this absorption feature reveals the high degree of light confinement within the structure. As a consequence this confinement—on the order of $\sim\lambda/10$—the device is capable of modulating light with an optical interrogation volume of as little as $\lambda^3/10$. (The actual modulation volume of our device was

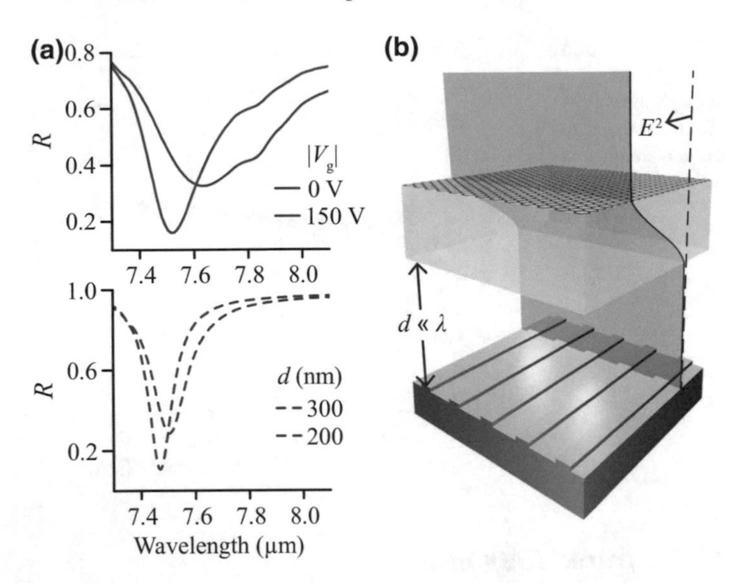

Fig. 5.7 Modulation and modelling of hexagonal boron nitride's Reststrahlen band. **a** Measured (upper panel) and modelled (lower panel) reflection spectra close to the TO phonon energy in hBN. The broader measured resonance width stems from the FTIR reflecting objective, which provides a range of measurement angles simultaneously ($\theta = 12 - 24°$) compared to the modelled $\theta = 15°$. **b** The air gap is significantly smaller than the mid-IR wavelengths, however, close to the TO phonon energy light is highly compressed within the hBN due to its large index of refraction

around λ^3 at mid-infrared frequencies. It can be reduced to the optical interrogation volume by using smaller suspended areas.)

As with shorter wavelengths, RCWA simulations agree very closely with the measured mid-infrared behaviour (Fig. 5.7a). The basic existence of the feature at $\lambda \sim 7.5$ μm is attributable to material absorption in hBN. Hence a similar reflection minimum also occurs in simulations of free standing thin hBN films, and can be modulated simply by changing the hBN thickness (for example from 110 to 140 nm). However, the hBN thickness is fixed experimentally. Instead we observe an analogous effect arising from the changing device structure. Altering d leads to changes in the optical field overlap with the hBN, since the field diminishes to zero at the gold surface (Fig. 5.7b). Although the nanostripe array was included in the above RCWA calculations, a much simpler Fresnel reflection-based model of a planar Au/air/hBN/graphene structure confirmed that it is not necessary for the observed mid-infrared modulation.

5.4 Modulation Frequency

5.4.1 Theoretical Modulation Frequency

To estimate the maximum modulation frequency of our device we approximate our device as a slab of hBN vibrating in one dimension with both ends fixed (Fig. 5.8). In our case $L \approx 50$ μm is the length of unfixed hBN that is free to vibrate and $d \approx 110$ nm is the thickness of the hBN. We shall take the hBN width b (going into the page) to be approximately equal to L. Euler-Bernoulli beam theory gives the fundamental frequency f of the simply supported hBN plate as [26]

$$f = \frac{\pi d}{2L^2}\sqrt{\frac{E}{12\rho}}, \tag{5.4}$$

where $E \approx 70$ GPa is the Young's modulus of the hBN and $\rho = 2.2$ g cm^{-3} is the density of hBN. This gives a fundamental vibration frequency (and hence possible modulation frequency) of the order of 100 kHz, which is comparable with the upper limits of other recently reported nanomechanical modulators [23]. It is worth noting that the dedicated design (for example using $L \approx 5$ μm and smaller hBN thickness) would allow one to increase this frequency to around 100 MHz.

5.4.2 Frequency Measurements

High speed experimental characterisation of the device was beyond the scope of this work, particularly as the device was not optimised for high frequency operation (e.g. we used graphene as one of the contacts, the area size was not optimised, etc.). However, we did measure the optical response of our modulator in the frequency range of 100 Hz–100 kHz available in our setup.

Figure 5.9 shows two examples of the operation of our modulator in reflection at a frequency of 1 kHz and 4 MHz. The green oscilloscope trace shows the applied ac gating voltage (which was superimposed with an offset gating voltage) and the yellow trace shows the modulated optical signal measured with a photodiode.

The typical frequency dependence of the optical modulation amplitude measured with a lock-in amplifier is given in Fig. 5.10a. We can see that the device still mod-

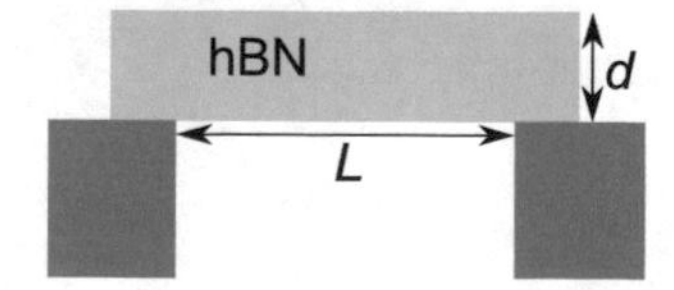

Fig. 5.8 Model of electro-mechanical modulator as vibrating suspended hBN slab. L is the length of unfixed hBN and d is its thickness

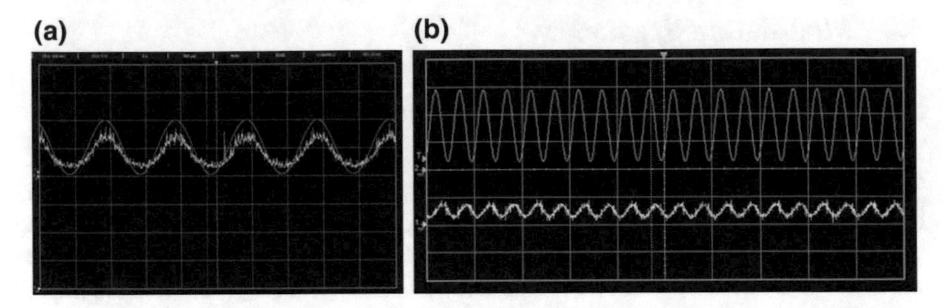

Fig. 5.9 Oscilloscope traces showing operation of modulator. Example of the operation of our modulator in reflection measured with a laser with wavelength 1.06 μm. The modulation frequency was **a** 1 kHz and **b** 4 MHz. The green oscilloscope trace shows the applied gating voltage and the yellow trace shows the modulated optical signal measured with a photodiode. In (**a**) the modulation is shown including the dc component of the reflected light, with origin at the bottom of the graph. In (**b**), the lower modulation amplitude necessitated the use of ac coupling in the oscilloscope

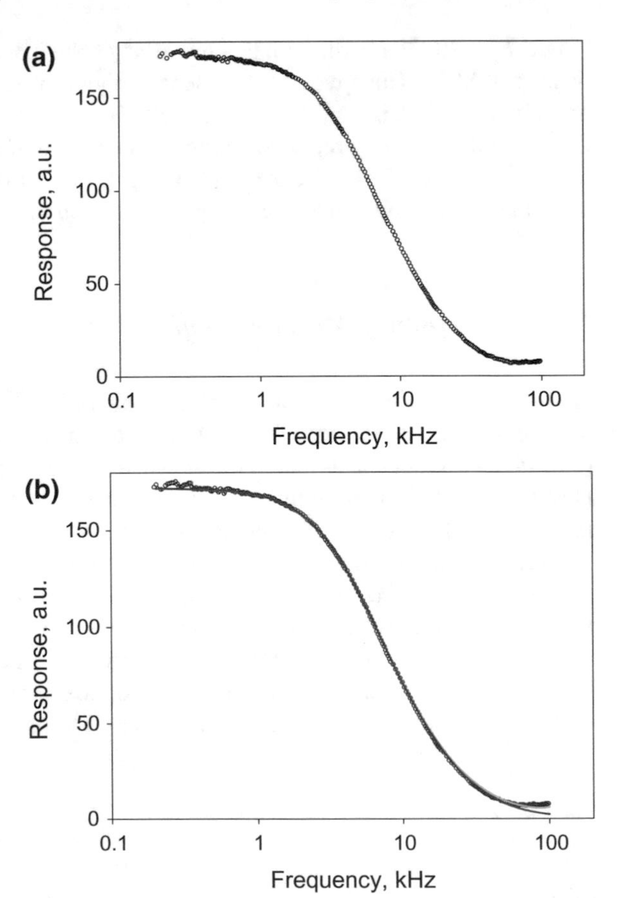

Fig. 5.10 Frequency dependence of modulation. **a** Typical measured frequency dependence of the optical modulation amplitude in our modulator. **b** The same frequency dependence plot as in (**a**) but with a fit to an overdamped response (red curve) and a fit as the product of a harmonic oscillator response and the overdamped electrical response (green curve)

ulates light at 100 kHz. Overall, it has an overdamped response (with damping frequency of 8 kHz) connected to the electrical part of our device (the inverse RC time constant for the gating voltage was estimated around 0.1 ms). The red curve of the plot in Fig. 5.10b shows the overdamped fit to the measured frequency dependence.

The fit is quite good except in the high frequency range approaching ∼100 kHz where we see a shallow peak of the experimental response (which is most probably connected with the mechanical resonance). In order to describe the combined electromechanical properties of the device we have fitted the measured data as the product of a harmonic oscillator response and the overdamped electrical response (green line in Fig. 5.9b), which fits the experimental data extremely well. This fit provides a mechanical resonance frequency of ∼120 kHz, which is in good agreement with estimate given above.

5.5 Discussion

One potential application of this modulation technique is the study of the nanomechanics of suspended 2D materials. Our modulator design could be adapted to detect the motion of a 2D heterostructure by measuring the dispersive coupling strength $g = \delta E/\delta z$, where δE is the shift in the collective plasmon energy and δz the displacement of the membrane. In the case of a plasmonic nanoarray supporting diffraction-coupled plasmon resonances with the 2D heterostructure suspended close to its surface this number was $g = 4$ meV nm^{-1} or 1000 GHz nm^{-1} which is quite large for a plasmonic device [24].

5.5.1 Maxwell Stresses

By treating our device as a parallel plate capacitor (with the graphene and gold nanoarray acting as the parallel plates; see Fig. 5.11), we estimate the Maxwell stresses experienced by the hBN and air within the device. $D_{a,b}$, $\epsilon_{a,b}$, $E_{a,b}$ and $d_{a,b}$ are respectively the displacement field, relative permittivity, electric field and thickness of the air gap (a) and the hBN (b) as defined in Fig. 5.11. We shall use $\epsilon_a = 1$, $\epsilon_b \approx 4$, $d_a \approx 100$ nm (when the applied voltage $V = 150$ V), and $d_b \approx 110$ nm. If the displacement field across the device is constant, then $D_a = D_b$ and $E_a = \frac{\epsilon_b}{\epsilon_a} E_b$. By finding the potential difference V across the device it can be shown that

$$E_b = \frac{V}{(d_b + \frac{\epsilon_b}{\epsilon_a} d_a)}. \quad (5.5)$$

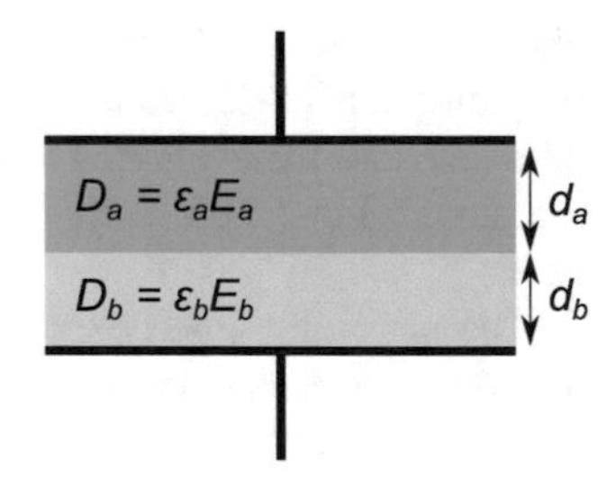

Fig. 5.11 Model of electro-mechanical modulator as a parallel-plate capacitor. $D_{a,b}$, $\epsilon_{a,b}$, $E_{a,b}$ and $d_{a,b}$ are respectively the displacement field, relative permittivity, electric field and thickness of the air gap (**a**) and the hBN (**b**)

An equivalent expression can be found for E_a. We can then use this value to calculate the Maxwell stress σ experienced by each dielectric [39],

$$\sigma_{a,b} = \epsilon_0 \epsilon_{a,b} E^2, \quad (5.6)$$

where $\epsilon_0 = 8.854 \times 10^{-12}$ F/m is the permittivity of free space. This yields a Maxwell pressure of around 20 atm (atmospheres) at the air-hBN interface at gating voltage of 100 V. The Grüneisen parameter is close to zero for most phonon modes in hBN, including the transverse optic phonon mode shown in Fig. 5.7 [40], meaning that the modulation of the absorption feature at $\sim$7.35 μm can only be attributed to the movement of the hBN flake.

5.5.2 *Further Device Optimisation*

Future nanomechanical modulators could benefit from decreasing the overall air gap thickness. Reducing the thickness of the air gap would also reduce the gate voltage required to produce the same Maxwell stress. Casimir forces could further enhance the modulation if the thickness of the hBN was decreased to bring the graphene-gold distance down to a few tens of nanometres [28]. In fact, RCWA simulations reveal that even the spectral positions of the Rayleigh resonances within our structure become increasingly sensitive to d as the air gap is reduced (Fig. 5.5). Reduction and optimisation of the structural dimensions of future devices could lead to further enhancement of modulation depths at all wavelengths.

5.6 Conclusion

We have shown that introducing an air gap of variable height into a graphene/hBN/nanoarray structure can allow for very strong optical reflection modulation via graphene gating. We have demonstrated that this effect can be used to create modulation from the mid-ultraviolet all the way to the upper Reststrahlen band of hBN (mid-infrared). The use of 2D atomic heterostructures allows for an extremely small optical interrogation volume of $\sim\lambda^3/10$. To see such strong effects over such a broad spectral range in a device controlled by graphene gating is unprecedented and opens up many exciting possibilities in the development of optical nanoelectromechanical systems.

References

1. K.S. Novoselov, A.K. Geim, S.V. Morozov, D. Jiang, M.I. Katsnelson, I.V. Grigorieva, S.V. Dubonos, A.A. Firsov, Two-dimensional gas of massless dirac fermions in graphene. Nature **438**(7065), 197–200 (2005)
2. F. Bonaccorso, Z. Sun, T. Hasan, A.C. Ferrari, Graphene photonics and optoelectronics. Nat. Photonics **4**(9), 611–622 (2010)
3. A.N. Grigorenko, M. Polini, K.S. Novoselov, Graphene plasmonics. Nat. Photonics **6**(11), 749–758 (2012)
4. W.L. Barnes, A. Dereux, T.W. Ebbesen, Surface plasmon subwavelength optics. Nature **424**(6950), 824–830 (2003)
5. V.G. Kravets, F. Schedin, A.N. Grigorenko, Extremely narrow plasmon resonances based on diffraction coupling of localized plasmons in arrays of metallic nanoparticles. Phys. Rev. Lett. **101**(8), 087403 (2008)
6. B. Auguié, W.L. Barnes, Collective resonances in gold nanoparticle arrays. Phys. Rev. Lett. **101**(14), 143902 (2008)
7. S. Zou, N. Janel, G.C. Schatz, Silver nanoparticle array structures that produce remarkably narrow plasmon lineshapes. J. Chem. Phys. **120**(23), 10871–10875 (2004)
8. D.K. Gramotnev, S.I. Bozhevolnyi, Plasmonics beyond the diffraction limit. Nat. Photonics **4**(2), 83–91 (2010)
9. V.G. Kravets, F. Schedin, R. Jalil, L. Britnell, R.V. Gorbachev, D. Ansell, B. Thackray, K.S. Novoselov, A.K. Geim, A.V. Kabashin, A.N. Grigorenko, Singular phase nano-optics in plasmonic metamaterials for label-free single-molecule detection. Nat. Mater. **12**(4), 304–309 (2013)
10. F.J. García de Abajo, Graphene nanophotonics. Science **339**(6122), 917–918 (2013)
11. R.R. Nair, P. Blake, A.N. Grigorenko, K.S. Novoselov, T.J. Booth, T. Stauber, N.M.R. Peres, A.K. Geim, Fine structure constant defines visual transparency of graphene. Science **320**(5881), 1308–1308 (2008)
12. F.H.L. Koppens, D.E. Chang, F.J. García de Abajo, Graphene plasmonics: a platform for strong light-matter interactions. Nano Lett. **11**(8), 3370–3377 (2011)
13. J. Chen, M. Badioli, P. Alonso-González, S. Thongrattanasiri, F. Huth, J. Osmond, M. Spasenović, A. Centeno, A. Pesquera, P. Godignon, A. Zurutuza Elorza, N. Camara, F.J. García de Abajo, R. Hillenbrand, F.H.L. Koppens, Optical nano-imaging of gate-tunable graphene plasmons. Nature **487**(7405), 77–81 (2012)
14. A.K. Geim, I.V. Grigorieva, Van der waals heterostructures. Nature **499**(7459), 419–425 (2013)
15. F. Withers, O. Del Pozo-Zamudio, A. Mishchenko, A.P. Rooney, A. Gholinia, K. Watanabe, T. Taniguchi, S.J. Haigh, A.K. Geim, A.I. Tartakovskii, K.S. Novoselov, Light-emitting diodes by band-structure engineering in van der waals heterostructures. Nat. Mater. **14**(3), 301–306 (2015)
16. L. Britnell, R.M. Ribeiro, A. Eckmann, R. Jalil, B.D. Belle, A. Mishchenko, Y.-J. Kim, R.V. Gorbachev, T. Georgiou, S.V. Morozov, A.N. Grigorenko, A.K. Geim, C. Casiraghi, A.N. Castro Neto, K.S. Novoselov, Strong light-matter interactions in heterostructures of atomically thin films. Science **340**(6138), 1311–1314 (2013)
17. T.J. Echtermeyer, L. Britnell, P.K. Jasnos, A. Lombardo, R.V. Gorbachev, A.N. Grigorenko, A.K. Geim, A.C. Ferrari, K.S. Novoselov, Strong plasmonic enhancement of photovoltage in graphene. Nat. Commun. **2**, 458 (2011)

18. F. Schedin, E. Lidorikis, A. Lombardo, V.G. Kravets, A.K. Geim, A.N. Grigorenko, K.S. Novoselov, A.C. Ferrari, Surface-enhanced raman spectroscopy of graphene. ACS Nano **4**(10), 5617–5626 (2010)
19. L. Ju, B. Geng, J. Horng, C. Girit, M. Martin, Z. Hao, H.A. Bechtel, X. Liang, A. Zettl, Y.R. Shen, F. Wang, Graphene plasmonics for tunable terahertz metamaterials. Nat. Nanotechnol. **6**(10), 630–634 (2011)
20. J. Kim, H. Son, D.J. Cho, B. Geng, W. Regan, S. Shi, K. Kim, A. Zettl, Y.R. Shen, F. Wang, Electrical control of optical plasmon resonance with graphene. Nano Lett. **12**(11), 5598–5602 (2012)
21. Y. Yao, M.A. Kats, R. Shankar, Y. Song, J. Kong, M. Loncar, F. Capasso, Wide wavelength tuning of optical antennas on graphene with nanosecond response time. Nano Lett. **14**(1), 214–219 (2014)
22. D. Ansell, I.P. Radko, Z. Han, F.J. Rodriguez, S.I. Bozhevolnyi, A.N. Grigorenko, Hybrid graphene plasmonic waveguide modulators. Nat. Commun. **6** (2015)
23. B.S. Dennis, M.I. Haftel, D.A. Czaplewski, D. Lopez, G. Blumberg, V.A. Aksyuk, Compact nanomechanical plasmonic phase modulators. Nat. Photonics **9**(4), 267–273 (2015)
24. A. Reserbat-Plantey, K.G. Schädler, L. Gaudreau, G. Navickaite, J. Güttinger, D. Chang, C. Toninelli, A. Bachtold, F.H.L. Koppens, Electromechanical control of nitrogen-vacancy defect emission using graphene nems. Nat. Commun. **7** (2016)
25. C.R. Dean, A.F. Young, I. Meric, C. Lee, L. Wang, S. Sorgenfrei, K. Watanabe, T. Taniguchi, P. Kim, K.L. Shepard, J. Hone, Boron nitride substrates for high-quality graphene electronics. Nat. Nanotechnol. **5**(10), 722–726 (2010)
26. D.J. Inman. *Engineering Vibration*, 4th edn. (Pearson, 2014)
27. Y.-N. Xu, W.Y. Ching, Calculation of ground-state and optical properties of boron nitrides in the hexagonal, cubic, and wurtzite structures. Phys. Rev. B **44**(15), 7787 (1991)
28. M. Bordag, I.V. Fialkovsky, D.M. Gitman, D.V. Vassilevich, Casimir interaction between a perfect conductor and graphene described by the dirac model. Phys. Rev. B **80**(24), 245406 (2009)
29. M. Wraback, H. Shen, S. Liang, C.R. Gorla, Y. Lu, High contrast, ultrafast optically addressed ultraviolet light modulator based upon optical anisotropy in ZnO films grown on R-plane sapphire. Appl. Phys. Lett. **74**(4), 507–509 (1999)
30. A.E. Oberhofer, J.F. Muth, M.A.L. Johnson, Z.Y. Chen, E.F. Fleet, G.D. Cooper, Planar gallium nitride ultraviolet optical modulator. Appl. Phys. Lett. **83**(14), 2748–2750 (2003)
31. R. Thijssen, E. Verhagen, T.J. Kippenberg, A. Polman, Plasmon nanomechanical coupling for nanoscale transduction. Nano Lett. **13**(7), 3293–3297 (2013)
32. N. Youngblood, Y. Anugrah, R. Ma, S.J. Koester, M. Li, Multifunctional graphene optical modulator and photodetector integrated on silicon waveguides. Nano Lett. **14**(5), 2741–2746 (2014)
33. W. Li, B. Chen, C. Meng, W. Fang, Y. Xiao, X. Li, Z. Hu, Y. Xu, L. Tong, H. Wang, W. Liu, J. Bao, Y. Ron Shen, Ultrafast all-optical graphene modulator. Nano Lett. **14**(2), 955–959 (2014)
34. J.P. Hugonin, P. Lalanne, *Reticolo Software for Grating Analysis* (2005)
35. M.G. Moharam, T.K. Gaylord, Rigorous coupled-wave analysis of metallic surface-relief gratings. JOSA A **3**(11), 1780–1787 (1986)
36. J.D. Caldwell, L. Lindsay, V. Giannini, I. Vurgaftman, T.L. Reinecke, S.A. Maier, O.J. Glembocki, Low-loss, infrared and terahertz nanophotonics using surface phonon polaritons. Nanophotonics **4**(1), 44–68 (2015)
37. N. Dabidian, I. Kholmanov, A.B. Khanikaev, K. Tatar, S. Trendafilov, S.H. Mousavi, C. Magnuson, R.S. Ruoff, G. Shvets, Electrical switching of infrared light using graphene integration with plasmonic fano resonant metasurfaces. ACS Photonics **2**(2), 216–227 (2015)
38. Y. Yao, M.A. Kats, P. Genevet, N. Yu, Y. Song, J. Kong, F. Capasso, Broad electrical tuning of graphene-loaded plasmonic antennas. Nano Lett. **13**(3), 1257–1264 (2013)
39. D.J. Griffiths, *Introduction to Electrodynamics* (1999)
40. G. Kern, G. Kresse, J. Hafner, Ab initio calculation of the lattice dynamics and phase diagram of boron nitride. Phys. Rev. B **59**(13), 8551 (1999)

Chapter 6
Strong Coupling of Diffraction Coupled Plasmons and Optical Waveguide Modes in Gold Stripe-Dielectric Nanostructures at Telecom Wavelengths

We propose a hybrid plasmonic device consisting of a planar dielectric waveguide covering a gold nanostripe array fabricated on a gold film and investigate its guiding properties at telecom wavelengths. The fundamental modes of a hybrid device and their dependence on the key geometric parameters are studied. A communication length of 250 μm was achieved for both the TM and TE guided modes at telecom wavelengths. Due to the difference between the TM and TE light propagation associated with the diffractive plasmon excitation, our waveguides provide polarisation separation. Our results suggest a practical way of fabricating metal-nanostripes-dielectric waveguides that can be used as essential elements in optoelectronic circuits.

The gold nanostripe arrays described in Chap. 4 were covered by a dielectric layer deposited by Vasyl Kravets. I characterised the samples using ellipsometry and Dmytro Kundys performed additional optical reflection measurements. I coauthored the manuscript with Vasyl Kravets and Alexander Grigorenko.

6.1 Introduction

Optical communication is the fastest means of information processing. However, conventional optical components are relatively bulky and cannot be packed together as compactly as the ubiquitous electronic components. Plasmonic nanostructures allow for waveguiding beyond the diffraction limit, leading to their consideration as potential replacements of electronic interconnects in the next generation of CMOS-integrated circuits [1, 2]. Several types of plasmonic waveguiding structures have been proposed and studied, including metal-insulator-metal and insulator-metal-insulator multilayered structures [3], strips [4], V-grooves [5], wedges [6] and nanoparticles chains [1]. The possibility of field localisation and enhancement

P. A. Thomas, *Narrow Plasmon Resonances in Hybrid Systems*,
Springer Theses, https://doi.org/10.1007/978-3-319-97526-9_6

suggests that plasmonic waveguides could have a large impact on applications at telecommunication and optical frequencies.

A unique property of these hybrid metallic waveguides is that they can simultaneously carry electronic and optical signals and therefore possess the potential to become electronically tuneable photonic devices, using graphene as a control element, for example [22]. However, such designs are limited by strong dissipative losses in the constituent plasmonic materials (usually noble metals such as gold or silver) [2, 5, 8–10]. Ohmic losses limit the propagation distance of highly-confined surface plasmon polaritons (SPPs) in planar waveguide structures to at best a few tens of micrometers [11]. Although it is possible to create waveguides in which SPPs can propagate for up to a few millimeters [4, 12], these long-range SPPs tend to be poorly confined [13]. For example, metal films and strips [1, 2, 13] can guide either long- or short-range SPPs by changing the film thickness, and decreasing the thickness of the film or strip results in poorer localisation of the long-range mode.

New waveguide designs have been investigated to overcome this limitation. V-grooves have previously been designed to increase the confined SPP propagation length to around 100 μm [5]. Metal strips and wedges are relatively easy to fabricate but are expected to exhibit relatively large propagation losses and may be sensitive to structural imperfections. A novel hybrid plasmonic waveguide that controls the coupling between dielectric and plasmonic modes has been suggested with high propagation length (approximately 40–150 μm) and field confinement [14]. Hybrid plasmonic waveguides in this case were formed by dielectric nanowires coupled to a metal surface which supports SPPs. The detailed experimental analysis of the actual impact of these factors on plasmon waveguides based on nanostripes is still to be performed. It has also been suggested that losses could be overcome by using gain-enhanced plasmonic metamaterials [8], replacing the currently used noble metals with other materials such as semiconductors [9], or by incorporating graphene into the waveguide structure [10]. However, these designs rely on complex fabrication techniques that in some cases have not yet been standardised.

In this work, we experimentally study hybrid plasmonic waveguides in a novel geometry where a dielectric layer is deposited directly onto a regular array of plasmonic nanostripes (NSs) fabricated on the surface of a gold film. This waveguide structure supports tightly-confined hybrid plasmon waveguide (HPWG) modes with long communication lengths (200–300 μm). We investigate the main properties of HPWG modes using angle-resolved optical measurements. The coupling between localised surface plasmon resonances (LSPRs) of the array and guided modes was controlled by tuning the geometrical parameters of gold nanostripes. Rabi splitting of HPWG TM modes was observed.

6.2 Sample Design

Electron-beam lithography was used to fabricate gold nanostripes on top of a flat metallic film. Figure 6.1a, b shows the schematic of the hybrid plasmon-waveguide system. Here, w and h are the width and height of Au NSs, respectively, d is the

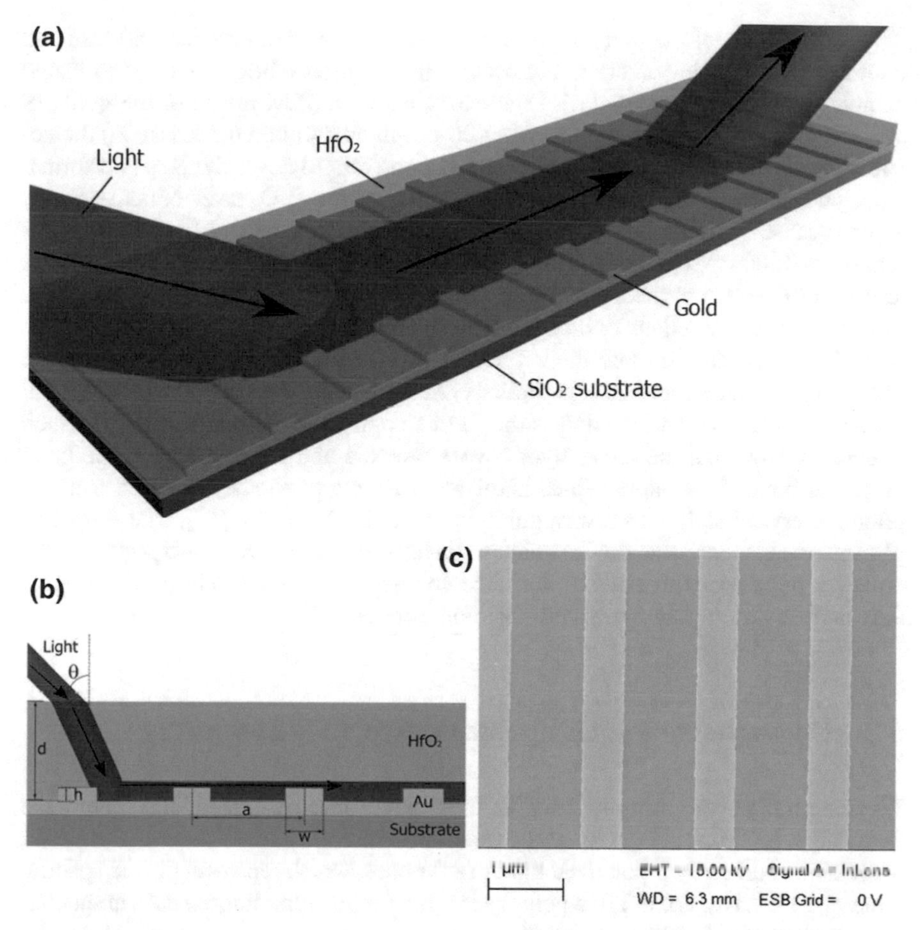

Fig. 6.1 Hybrid plasmonic waveguide devices. **a** A schematic illustration of the hybrid structure illuminated by a light with equal incident and scattering angles. **b** Schematic illustration of trapping of light (at incident angle θ) in NSs-waveguide layer due to scattering on the nanostructure and internal reflection. **c** SEM image of the gold nanostripe sample without the waveguide layer

thickness of the top waveguide layer and a is the period of the NS array. Samples were fabricated using standard electron-beam lithography techniques; see Sect. 2.3.2. To choose the optimal height of gold NSs it is possible to simulate our nanostructure using the analytical theory described in [15] and experimentally confirmed for terahertz surface plasmon waveguides in [16]. We have investigated the diffraction coupled plasmon resonances in pure gold NSs with fixed stripe wide 400 nm and height in the range from 60 to 90 nm with step of 10 nm. It was found that the stripe height of 70 nm is optimal for achieving the sharpest plasmonic resonances for telecom wavelengths of the best quality: $Q = \lambda_R/\Delta\lambda$ (where λ_R is the spectral position of plasmon resonance and $\Delta\lambda$ is the full-width at half-maximum); see Chap. 4 for further details.

The period, a, of the line arrays on each sample was fixed around 1500 nm and confirmed by scanning electron microscopy (SEM) after fabrication (we fabricated samples with different w and a). Figure 6.1c shows a SEM image of the gold NS sample without the waveguide layer. A 400 nm-thick hafnium oxide (HfO_2) dielectric layer was deposited by electron beam evaporation to cover the NSs and form a waveguide layer on top of the samples. We have chosen HfO_2 as an optical guiding layer because of its chemical stability and high refractive index of ~1.9 (at wavelengths of 1300–1550 nm). For HfO_2, the trapping angle of light can be evaluated as arcsin $(1/n) \approx 32°$ at the air-film interface. This implies that only radiation going into the top 1.1 steradians (which is about $1.1/(4\pi) \approx 8\%$ of the total angle) will partially escape the film into the air. There exists an optimum combination of the gold stripe width, w, and their periodicity, a, in the dielectric layer (with high real refractive index, n) that maximises the light trapping and optical guiding for such structures. Note that the HfO_2 layer covers the NSs and forms a waveguide layer on top of the entire sample, which is different from a previously reported metallic photonic crystal slab with a waveguide beneath the NS array [17]. The proposed plasmon-waveguide structure is similar to the structure recently investigated in [18]; however, our devices are adjusted for telecom frequencies and can be used as important elements for highly-integrated photonic circuits.

6.3 Characterisation Using Spectroscopic Ellipsometry

We characterised our samples using variable angle spectroscopic ellipsometry; see Sect. 2.2.5. Note that the ellipsometric function ψ simultaneously describes plasmon resonances excited by p-polarized light (TM modes, which represent dips in spectra, when r_p tends to zero) and TE (s-polarized) modes (corresponding to peaks in spectra, when values of r_s become very small).

The typical high quality plasmonic resonance based on diffraction coupled plasmons measured in our samples without a waveguide layer is shown in Fig. 6.2a. The sample has lines $w = 410$ nm wide (with a period $a \approx 1500$ nm fabricated on a 65 nm gold film.) One can see that a very narrow resonance peak at telecom wavelengths (~1300–1550 nm) was observed as the incident angle θ increases from 45° to 70°. Rayleigh-like anomalies in pure plasmonic gold NSs have to be considered as the physical origin for these spectral features [19]. Remarkably, the bandwidth of the resonance at the incident angle of 70° is only ~8 nm which is approximately 1/200 of the central wavelength 1500 nm. The origin of extremely narrow plasmon resonances in gold stripe arrays fabricated on a gold film is discussed in Chap. 4. The drops in the ellipsometric parameter ψ (Fig. 6.2a) and reflection of p-polarized light around the telecom wavelengths were dependent on the angle of incidence. The minimal half-width of the resonant feature observed was just 5 nm (Fig. 4.3a) which is much smaller than typical values for LSPRs. Extremely small widths of resonances in reflection are due to the occurrence of diffraction-coupled localised

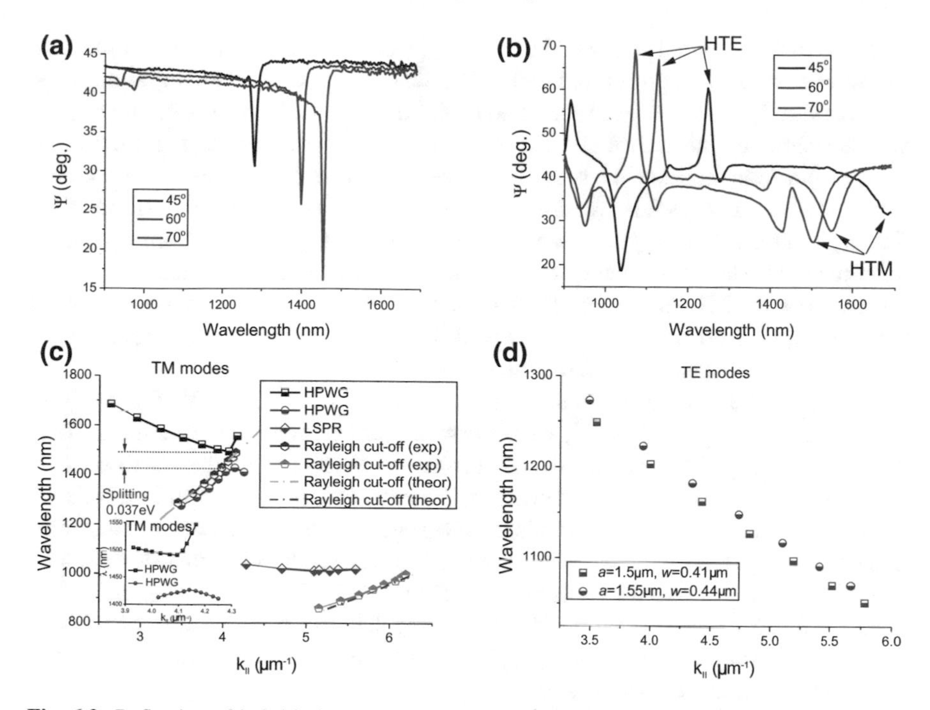

Fig. 6.2 Reflection of hybrid plasmon waveguides and properties of hybrid plasmon waveguide modes. **a** Ellipsometric spectra ψ showing the collective plasmon resonances of pure gold stripes fabricated on a gold layer. **b** Ellipsometric spectra of fully fabricated device with identification of HPWG TE and TM modes. **c**, **d** Experimental dispersion of guided modes as a function of the wavevector $k_{||}$: (**c**) TM modes (inset demonstrates in detail the dispersion curves near the splitting region); (**d**) TE modes. The data are measured for the sample with $w = 410$ nm, $a = 1500$ nm, $h = 70$ nm and $d = 400$ nm

plasmon resonances [19] and the drop of light intensity down to zero is due to a strong enhancement of local electric fields near the nanostripes.

Next, we covered the Au NS arrays with a waveguide layer made of HfO_2. The waveguide layer supports propagating waveguide modes coupled to incident light by the NS array. These modes are often referred to as hybrid plasmon waveguide modes and are quasiguided [17]. They manifest themselves as drops in reflection from the structure. Figure 6.2b–d shows the experimental function ψ for the same sample with a 400 nm-thick waveguide layer deposited on top. (The fundamental TM and TE HPWG modes are marked by the text.) Compared with the spectral dependence shown in Fig. 6.2a, the main resonance wavelength that corresponds to p-polarisation shifts from 1450 nm (for $\theta = 70°$) to 1505 nm; see Fig. 6.2b. The minimum at $\sim$1500 nm splits into two dips and a new peak appears at 1075 nm. Interestingly, the peak spectral position for the sample with the waveguide layer demonstrates blueshift with increasing angle of incidence, which is opposite to the redshift observed for the sample without the waveguide layer; compare Fig. 6.2a, b.

In Fig. 6.2c we plot the resonant wavelengths of HPWG TM modes as a function of the wavevector component parallel to the wavevector of the stripe grating, $k_{\|} = k_0 \sin\theta$, with k_0 the free space vector and θ the angle of incidence. For completeness, we also show the theoretical and experimental dispersion curve of the diffraction-coupled plasmon modes of the bare NS arrays without the HfO_2 layer. (These modes correspond to the Rayleigh-Wood anomalies [19–21].) The expected position of the Rayleigh cut-off wavelength for air is $\lambda_R = \frac{a}{m}(1 + \sin\theta)$, which gives very good agreement with the experimental data for $m = 2,3$, the dash-point lines in Fig. 6.2c, measured on the NS array before HfO_2 deposition. Note that the dispersion curve of the Rayleigh cut-off wavelength for $m = 2$ lies exactly between two guided modes denoted by black squares and red circles in Fig. 6.2c.

The HPWG TM mode exhibits a splitting energy of ~0.037 eV (Fig. 6.2c). In the case of large $k_{\|} > 4.3\mu\text{m}^{-1}$, we observe an almost non-dispersive mode (blue diamonds at Fig. 6.2c) which is most probably connected to the localised out-of-plane resonance of a gold nanostripe. To characterise the coupling efficiency between incident light and plasmonic and waveguide modes, we consider the splitting energy (Fig. 6.2b, c) and the full width of resonance. The splitting was estimated around $E_{sp} \sim 0.037$ eV (see above) while the linewidth of resonance is smaller than $\Delta E_{res} \sim 0.025$ eV (a dip at $\lambda \sim 1500$ nm in Fig. 6.2b for $\theta = 70°$ was taken as an example). Due to fact that the energy splitting is larger than the resonance linewidth, we conclude that the coupling is strong.

In the case of TE modes, gold NSs only function as periodical scatters that do not support LSPRs. Figure 6.2d shows the measured spectral dependence of HPWG TE mode. As the periodicity of NSs, a, increases from 1500 to 1550 nm, the measured reflection peak shifts toward larger wavelengths (Fig. 6.2d). Finite element modelling confirms the plasmonic origin of the guided modes; see Fig. 6.3.

6.4 Communication Length of Hybrid Plasmon-Wave-Guide System

We now describe the most important result of this study. We found that the hybrid plasmon-waveguide system can be used in plasmonic devices that require light transfer over large communication lengths (hundreds of micrometers). The communication length of a waveguide is the distance between the probing input and gathered output signal. An analogous conclusion has been reached for a structure of inverted geometry with surface plasmon polaritons excited at a gold surface with a buried grating in the Kretschmann geometry [22].

Figures 6.1a and 6.4 illustrate the schematic of the experiment where light is incident on the device under an oblique angle, propagates along the nanostructure as a HPWG mode and is re-emitted at the reflection angle. We have measured the spatial distance between the input and output spot of the incident-reflected light using an optical microscope (see below). It was found that the hybrid plasmon-guided modes

Fig. 6.3 Finite element simulation of electric fields in a hybrid air-dielectric-metal stripe waveguide. Finite-element simulations were carried out using HFSS for light incident on the waveguide at 70° with periodic boundary conditions. The simulated fields clearly show coupling of the fields at the corner of the gold nanostripe associated with localised surface plasmon resonances with the fields corresponding to propagating modes away from the gold nanostripe. Simulations also showed this device has a plasmon resonance wavelength at ~1.5 μm (graph not shown), agreeing with out experimental observations in Figs. 6.2 and 6.5

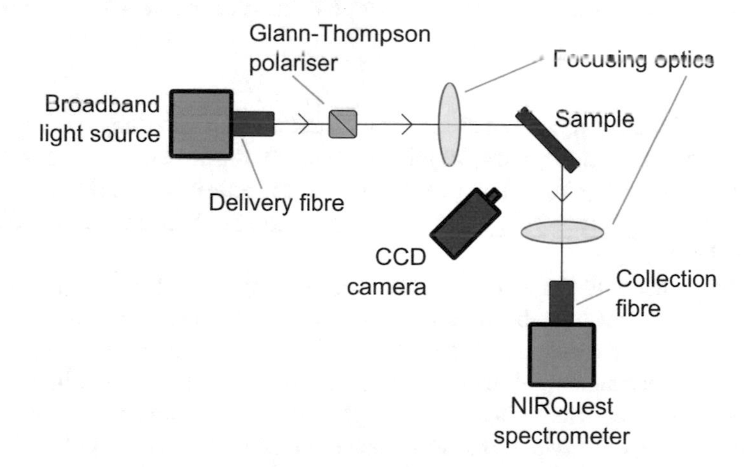

Fig. 6.4 Optical setup for communication length measurements

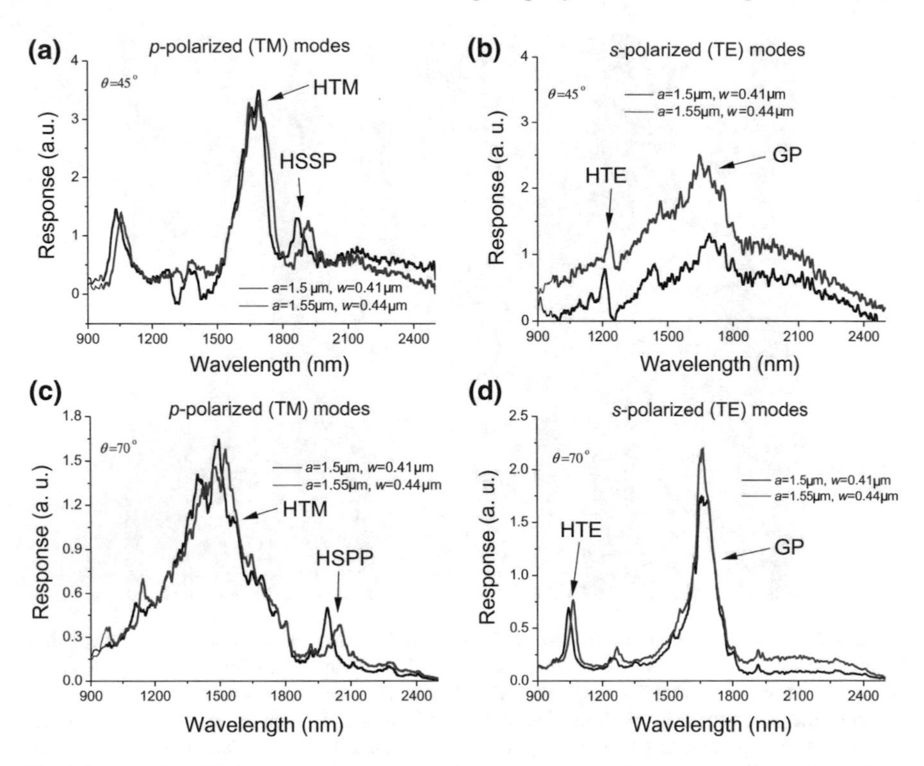

Fig. 6.5 Transfer of light along hybrid plasmonic waveguides. **a**, **b** The spectral dependence of the transmitted light intensity inside hybrid plasmonic-waveguide at long distance ($\sim$250 μm) for TM (**a**) and TE (**b**) modes at angle of incidence $\theta = 45°$. **c**, **d** The same case as (**a**, **b**) but for an angle of incidence $\theta = 70°$

can transmit photon energy at the distance of at least 250 μm. This length is larger than the surface plasmon propagation lengths in comparable waveguide designs (which tend to be around 100 μm) and is comparable with the upper limit of the propagation length of surface plasmon polaritons in a flat gold film at 1550 nm (calculated to be 730 μm [22, 23]).

The experiments to determine communication length for both the TM and TE hybrid guided modes have been carried out using a confocal microscope with the ability to accurately excite and collect the light at various spatial positions on the sample. The experimental setup consisted of a laser driven broadband light source, Glan-Thompson polariser on a rotating stage and focusing optics (x20 super long working distance objective, working distance 25 mm). We have used multimode broadband fibres with core diameter of 20 μm to deliver and collect the light signal, which acted as a spatial filter. The plasmon-waveguide structure was illuminated on the left edge (see Fig. 6.1a) with a broad spectrum light at angles of incidence 45° and 70° by using an objective lens to give a spot size less than 30 μm. To image the focused light spot we have used both an 8% reflectivity beam splitter and a CCD camera in a standard microPL arrangement or high zoom, high magnification CCD camera mounted at a normal incidence to the sample. A small fraction of the excited light was backscattered by the sample grating enabling precise monitoring and positioning of the focused spot. The sample was mounted on a 3D micro-positioning stage.

Before each measurement (i.e. 45° and 70°) the sample was replaced by a silver mirror in order to take a reference of the reflected light in each polarisation. In order to define our out-coupling position in space (the other end of the communication length) we have used confocal reverse light coupling. This is achieved when the light is coupled from the collection fibre port and is focused by a collection lens forming a spot on the other end of the sample. The position of this spot was carefully adjusted by moving the XYZ translation of the collection lens. The collection spot could also be viewed by the CCD cameras.

Once the alignment of both focusing and collection points was completed the light was again coupled from the input fibre port. The collection fibre was sent to a NIRQuest spectrometer to analyse the light output collected only from a predefined spot on the sample. The spatial distance between the input and output spot of the incident-reflected light was estimated using an optical microscope. In our experiment, we have measured the spectral dependence of the transmitted light at long spatially distributed distance of 250 μm for our plasmon-waveguide structure (Fig. 6.1). To check the optical setup, we have measured transmitted light intensity in same geometry as shown in Fig. 6.1a, b for pure gold stripes and gold stripes with 70 nm-thick HfO_2 film for the same communication length. In both cases the intensity of transmitted light was at level of standard noise (measurements not shown here).

Figure 6.5 shows the spectral dependence of the intensity of light transmitted at a distance of 250 μm for two values of NSs periodicity, a, for p- and s- polarised light when the angle of incidence is equal to 45° (smaller than the Brewster angle, $\theta_B \approx 61°$, for HfO_2) and 70° (larger than the Brewster angle for HfO_2). The light transmitted along the waveguide modes correspond to peaks in measured transmission. (For clarity, we have labelled the transmitted modes in the Discussion section.) This yields a length for the propagation of HPWG modes on the order of hundreds of micrometres, which suggests a strong contribution from coupling between plasmon and waveguide modes. While the detailed mode analysis of the fabricated plasmon-waveguide devices and their waveguiding characteristics and the excitation efficiency is still to be carried out, radiation transfer between the in- and out-coupling gold stripes is clearly seen in Fig. 6.5. In addition, the position of propagated modes can be tuned in the plasmon-waveguide device by changing the angle of incidence of pumping light. The excitation conditions for HPWG TE and TM modes are different which implies that the structures can also function as a polarisation sensitive device.

6.5 Discussion

A flat dielectric layer of HfO_2 deposited on a gold film possesses propagating waveguide modes. In the presence of a nanostripe array fabricated on a gold film, these waveguide modes can couple to the light continuum outside the waveguide and become leaky and lossy. The coupling happens due to re-emission of light by the nanostripes (diffraction coupling with array wavevector mismatch). This coupling is most effective for wavelengths close to the localised plasmon resonance of a nanostripe. As a result, localised plasmon resonances in the studied structure can be important for device operation since they define both coupling strength and the mode loss coefficient.

In the first approximation, we model our device as a metal-dielectric-air multilayer system. In such a system the dispersion relation for TM and TE-guided modes can be found from the equations [24, 25]

$$\tan(kd) = \frac{\epsilon_2 k(\epsilon(\omega)\gamma + \epsilon_0\delta)}{\epsilon_0\epsilon(\omega)k^2 - \epsilon_2^2\gamma\delta} \quad \text{(TM modes)} \tag{6.1a}$$

$$\tan(kd) = \frac{k(\gamma+\delta)}{k^2 - \gamma\delta} \quad \text{(TE modes)} \tag{6.1b}$$

where ϵ_0, ϵ_2 and $\epsilon(\omega)$ are the dielectric constants for air, HfO_2 and gold. The parameters k, γ, and δ are given by [24, 25]:

$$k = \sqrt{\epsilon_2 k_0^2 - \beta^2} \tag{6.2}$$

$$\gamma = \sqrt{\beta^2 - \epsilon_0 k_0^2} \tag{6.3}$$

$$\delta = \sqrt{\beta^2 - \epsilon(\omega)k_0^2}, \tag{6.4}$$

where β is the wavevector of the guided modes and $k_0 = 2\pi/\lambda$ is the wavevector of photons in air. Figure 6.6a plots the spectral dependence of the wavevector of fundamental (lowest energy) TE and TM modes calculated with the help of Eqs. 6.1a and 6.1b. These wavevectors are large and cannot be produced by external light. However, in the presence of a periodic NS array, the momentum mismatch can be overcome by the reciprocal vector of the array $G = 2\pi/a$. Figure 6.6a shows that the line $k_\parallel + G$ (plotted for the angle of incidence $\theta = 45°$ as an example) indeed intersects the TE and TM mode lines at the wavelengths 1340 nm and 1650 nm respectively, which correspond to the excitation conditions of hybrid plasmon-waveguide modes (the intersections are shown by open circles). These theoretical values are close to the experimentally observed HPWG TE and TM modes in both reflection (at wavelengths of 1250 nm for HPWG TE mode and 1680 nm for TM mode; see Fig. 6.2b) and light transmitted along the waveguide (1210 nm for HPWG TE mode and 1690 nm for TE mode; Fig. 6.5a, b). By changing the angle of incidence we can find spectral behaviour of HPWG modes and compare it with the measured data; see Fig. 6.5b. It is clear that calculated fundamental HPWG TE and TM modes (maroon and brown lines) are in reasonably good agreement with the measured positions of HPWG TE and TM modes (maroon and brown circles).

In addition to HPWG modes, our structures possess other plasmon modes. First, surface plasmon polaritons can be excited at the metal interface. A SPP mode on a flat gold film is described by the following spectral relation

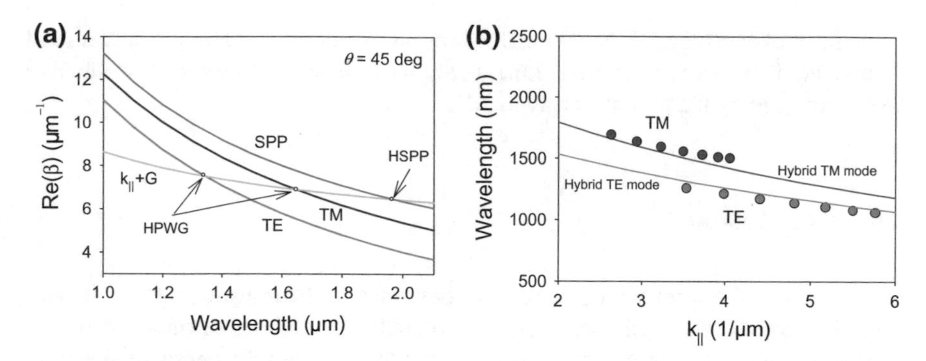

Fig. 6.6 Hybrid plasmon modes of the structure. **a** The dispersion of TE and TM modes of a flat dielectric waveguide (air-dielectric-metal) calculated for angle of incidence $\theta = 45°$, maroon and brown curves, respectively, combined with the dispersion of the SPP mode (dielectric-metal) and the light line with momentum mismatch. **b** Comparison of calculated dispersion of HPWG TE and TM modes with experimental data plotted in Fig. 6.2c, d

$$k(\omega) = \frac{\omega}{c}\sqrt{\frac{\epsilon(\omega)\cdot\epsilon_2}{\epsilon(\omega)+\epsilon_2}}, \tag{6.5}$$

where $k(\omega)$ is the propagating vector of the SPP mode. (We previously derived this relation in Sect. 2.2.1; see Eq. 2.30.) Fig. 6.6a plots the SPP mode (green line). As in case of waveguide TE and TM modes, SPP modes can only be excited by external light due to the presence of the periodic NS array. The intersection of the line $k_{\|} + G$ with the SPP line gives the position of the hybrid surface plasmon polariton mode (HSPP) at wavelength 1970 nm, which is reasonably close to the measured position of HSPP (at a wavelength of 1860 nm; see Fig. 6.5a). By comparing the amount of light transmitted by HSPP and HPWG modes, we can conclude that HPWG modes transfer about 4 times more light than HSPP at the communication distance of 250 μm. Second, localised out-of-plane plasmons (excited in p-polarisation) can re-scatter light to reach the detector. These plasmons result in a weakly dispersive mode seen in Fig. 6.2c. Third, in addition to HPWG TE modes in s-polarisation we observe an "anomalous resonance" (AR) at around 1750 nm which is vaguely independent of angle of incidence (see Fig. 6.5b, d). The origin of this feature requires further investigation and will be discussed in future work.

It is interesting to note that the band gap observed for the fundamental TM mode (p-polarisation) is opened at the propagating wavevector $k_{\|} \approx G$, which is different from the standard Bragg condition $k_{\|} = G/2$ and corresponds to higher order Bragg matching. Near the bandgap position, one can expect slow group velocity of the light. The group velocity of propagating light modes can be found as $v_g = \mathrm{d}\omega/\mathrm{d}\beta$. It corresponds to the group effective refractive index $n_g = c/v_g = \mathrm{d}\beta/\mathrm{d}\omega$. In case of our structures, $\beta = k_{\|} + G$ and the corresponding group index can be approximately evaluated from the relation $n_g = c \cdot \Delta k_{\|}/\Delta\omega$ which yields $n_g \approx 10$ from data plotted in the inset of Fig. 6.2c. We conclude this section by noting that HPWG modes (and others) require a more advanced treatment which would take into account an effective layer produced by the periodic array of NS.

6.6 Conclusion

Hybrid metal-dielectric waveguides have been shown to transfer light with relatively low losses and large communication distances (hundreds of micrometres) at telecommunication wavelengths. Based on photonic waveguide theory, the location and bandwidth of the hybrid plasmonic-waveguide resonances can be tuned by the geometrical parameters of the gold NSs and the angle of incidence of light. As a result, metal-nanostripes-dielectric waveguides could be applied to process information in optoelectronic circuits.

References

1. S.I. Bozhevolnyi, *Plasmonic Nano-Guides and Circuits* (Pan Stanford, 2008)
2. D.K. Gramotnev, S.I. Bozhevolnyi, Plasmonics beyond the diffraction limit. Nat. Photonics **4**(2), 83–91 (2010)
3. R. Zia, M.D. Selker, P.B. Catrysse, M.L. Brongersma, Geometries and materials for subwavelength surface plasmon modes. JOSA A **21**(12), 2442–2446 (2004)
4. A. Boltasseva, T. Nikolajsen, K. Leosson, K. Kjaer, M.S. Larsen, S.I. Bozhevolnyi, Integrated optical components utilizing long-range surface plasmon polaritons. J. Lightwave Technol. **23**(1), 413 (2005)
5. S.I. Bozhevolnyi, V.S. Volkov, E. Devaux, T.W. Ebbesen, Channel plasmon-polariton guiding by subwavelength metal grooves. Phys. Rev. Lett. **95**(4), 046802 (2005)
6. E. Moreno, S.G. Rodrigo, S.I. Bozhevolnyi, L. Martín-Moreno, F.J. Garcia-Vidal, Guiding and focusing of electromagnetic fields with wedge plasmon polaritons. Phys. Rev. Lett. **100**(2), 023901 (2008)
7. D. Ansell, I.P. Radko, Z. Han, F.J. Rodriguez, S.I. Bozhevolnyi, A.N. Grigorenko, Hybrid graphene plasmonic waveguide modulators. Nat. Commun. **6** (2015)
8. O. Hess, J.B. Pendry, S.A. Maier, R.F. Oulton, J.M. Hamm, K.L. Tsakmakidis, Active nanoplasmonic metamaterials. Nat. Mater. **11**(7), 573–584 (2012)
9. P.R. West, S. Ishii, G.V. Naik, N.K. Emani, V.M. Shalaev, A. Boltasseva, Searching for better plasmonic materials. Laser Photonics Rev. **4**(6), 795–808 (2010)
10. V.G. Kravets, R. Jalil, Y.-J. Kim, D. Ansell, D.E. Aznakayeva, B. Thackray, L. Britnell, B.D. Belle, F. Withers, I.P. Radko, Z. Han, S.I. Bozhevolnyi, K.S. Novoselov, A.K. Geim, A.N. Grigorenko, Graphene-protected copper and silver plasmonics. Sci. Rep. **4** (2014)
11. S.A. Maier, *Plasmonics: Fundamentals and Applications* (Springer Science & Business Media, 2007)
12. R. Charbonneau, P. Berini, E. Berolo, E. Lisicka-Shrzek, Experimental observation of plasmon-polariton waves supported by a thin metal film of finite width. Opt. Lett. **25**(11), 844–846 (2000)
13. P. Berini, Plasmon-polariton waves guided by thin lossy metal films of finite width: bound modes of symmetric structures. Phys. Rev. B **61**(15), 10484 (2000)
14. R.F. Oulton, V.J. Sorger, D.A. Genov, D.F.P. Pile, X. Zhang, A hybrid plasmonic waveguide for subwavelength confinement and long-range propagation. Nat. Photonics **2**(8), 496–500 (2008)

15. F.J. Garcia-Vidal, L. Martin-Moreno, J.B. Pendry, Surfaces with holes in them: new plasmonic metamaterials. J. Opt. A Pure Appl. Opt. **7**(2), S97 (2005)
16. G. Kumar, S. Li, M.M. Jadidi, T.E. Murphy, Terahertz surface plasmon waveguide based on a one-dimensional array of silicon pillars. New J. Phys. **15**(8), 085031 (2013)
17. A. Christ, S.G. Tikhodeev, N.A. Gippius, J. Kuhl, H. Giessen, Waveguide-plasmon polaritons: strong coupling of photonic and electronic resonances in a metallic photonic crystal slab. Phys. Rev. Lett. **91**(18), 183901 (2003)
18. J. Zhang, W. Bai, L. Cai, X. Chen, G. Song, Q. Gan, Omnidirectional absorption enhancement in hybrid waveguide-plasmon system. Appl. Phys. Lett. **98**(26), 261101 (2011)
19. V.G. Kravets, F. Schedin, A.N. Grigorenko, Extremely narrow plasmon resonances based on diffraction coupling of localized plasmons in arrays of metallic nanoparticles. Phys. Rev. Lett. **101**(8), 087403 (2008)
20. V.G. Kravets, F. Schedin, G. Pisano, B. Thackray, P.A. Thomas, A.N. Grigorenko, Nanoparticle arrays: from magnetic response to coupled plasmon resonances. Phys. Rev. B **90**(12), 125445 (2014)
21. L. Rayleigh, On the dynamical theory of gratings. Proc. Roy. Soc. Lond. Ser. A Contain. Pap. Math. Phys. Character **79**(532), 399–416 (1907)
22. J. Jose, F.B. Segerink, J.P. Korterik, A. Gomez-Casado, J. Huskens, J.L. Herek, H.L. Offerhaus, Enhanced surface plasmon polariton propagation length using a buried metal grating. J. Appl. Phys. **109**(6), 064906 (2011)
23. J. Zhang, L. Cai, W. Bai, G. Song, Hybrid waveguide-plasmon resonances in gold pillar arrays on top of a dielectric waveguide. Opt. Lett. **35**(20), 3408–3410 (2010)
24. I.P. Kaminow, W.L. Mammel, H.P. Weber, Metal-clad optical waveguides: analytical and experimental study. Appl. Opt. **13**(2), 396–405 (1974)
25. S.C. Rashleigh, Four-layer metal-clad thin film optical waveguides. Opt. Quantum Electron. **8**(1), 49–60 (1976)

Chapter 7
Phase-Sensitive Detection of HT-2 Mycotoxin Using Graphene-Protected Copper Plasmonics

Surface plasmon resonance (SPR) biosensing has been a commercially established sensing technique for over a decade. However, commercial systems, which measure shifts in the position of SPRs in gold thin films, struggle to detect lower concentrations of smaller molecules. After an extensive search, graphene-protected copper thin films have emerged as a viable alternative to gold in SPR biosensing applications. Copper thin films possess stronger, higher-quality plasmon resonances than their gold counterparts with resonance minima close to zero in amplitude, giving rise to enhanced phase sensitivity. Transferring a single layer of graphene on top of the copper protects the copper film from oxidation, preserving the quality of the plasmon resonances and allowing for surface functionalisation.

We present results from a feasibility study of phase-sensitive graphene-protected copper SPR biosensing. We have detected HT-2 mycotoxin, a small (mass $\approx$ 424 Da) molecule important in food safety applications. Commercial SPR systems have been shown to detect HT-2 mycotoxin using a specific antibody-based assay, but only down to concentrations of 10 ng/mL and with the use of a secondary antibody binding to the first antibody-HT-2 mycotoxin conjugate. We have detected HT-2 mycotoxin at concentrations as low as 1 pg/mL using graphene-protected copper using the same primary antibody but without the use of the secondary antibody. Additionally, we have shown that the sensitivity of our assay using phase measurements is around 6 fg/mL, with an areal mass sensitivity below 50 ag/mm^2, corresponding to the detection of tens of molecules. Our results demonstrate a simple, practical and relatively cheap way of enhancing the sensitivity of SPR biosensing by at least 4 orders of magnitude.

In this work I prepared and characterised samples with assistance from Fan Wu and Vasyl Kravets. The surface functionalisation protocol was developed by performed by Henri Arola and Miika Soikkeli. I conducted biosensing experiments with assistance from all other authors. Analysis of results was carried out with Fan Wu. I wrote the manuscript and prepared all figures, except for Fig. 7.8 which was prepared by Henri Arola and Miika Soikkeli.

P. A. Thomas, *Narrow Plasmon Resonances in Hybrid Systems*,
Springer Theses, https://doi.org/10.1007/978-3-319-97526-9_7

7.1 Introduction

After decades of investigation, biosensing remains the most widespread application of surface plasmons [1–4]. As previously outlined in Sect. 2.2.3, surface plasmon resonances are exceptionally sensitive to the dielectric environment close to the surface of the plasmonic nanostructure. Functionalising the surface of a plasmonic nanostructure with receptors (usually antibodies or aptamers) allows for specific molecules to bind near the surface. The binding shifts the position of the surface plasmon resonance which can be measured either by comparing two reflection spectra or, more simply, by measuring the change in intensity of reflected light at a wavelength close to the plasmon resonance wavelength [1]. A schematic of a typical SPR biosensing experiment is presented in Fig. 7.1.

One of the major advantages of SPR biosensing is that it is a label-free detection technique [4]: that is, labelling of the target molecules (for example, with fluorochromes) is not necessary. SPR biosensing is therefore a relatively straightforward sensing technique which does not require complicated sample preparation.

The past decade has seen a large research effort aimed at improving the sensitivity of SPR biosensing. These efforts have largely focused on the use of gold nanoparticles, either in self-assembled [5] or lithographically designed [6, 7] arrays. Such studies have yielded impressive results but any potential for commercialisation is usually limited by cost. SPR biosensing utilises the Kretschmann configuration (Sect. 2.2.2) with a thin gold film, meaning that existing commercial SPR systems are already very expensive [8].

A simple way of improving the sensitivity of SPR biosensing without making the technique prohibitively expensive (or potentially even cheaper) would be to replace the gold thin film with the thin film of another, cheaper metal that possess stronger plasmon resonances [9]. Such metals, such as copper and silver, exist, but have so far been limited in their application to SPR by their relative chemical instability. It has, however, been shown that it is possible to preserve the quality of the plasmon resonances in copper and silver thin films by using graphene as a protective barrier on the surface of the metal [10]. Typical surface plasmon resonances of gold and graphene-protected copper in the Kretschmann configuration are shown in Fig. 7.2.

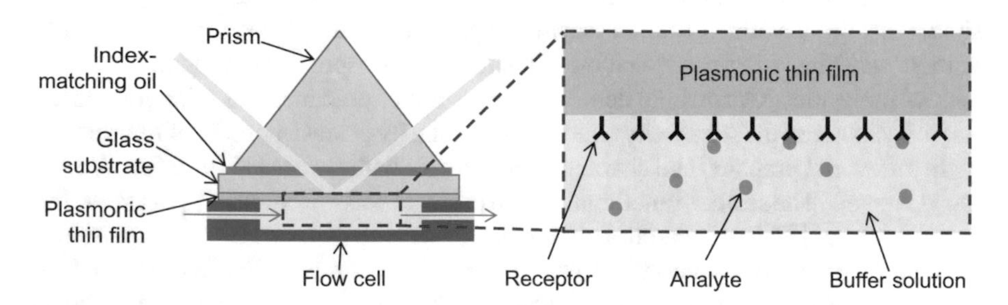

Fig. 7.1 Left: Kretschmann SPR configuration for biosensing. Right: Principle of binding of analytes at the surface of the plasmonic metal film

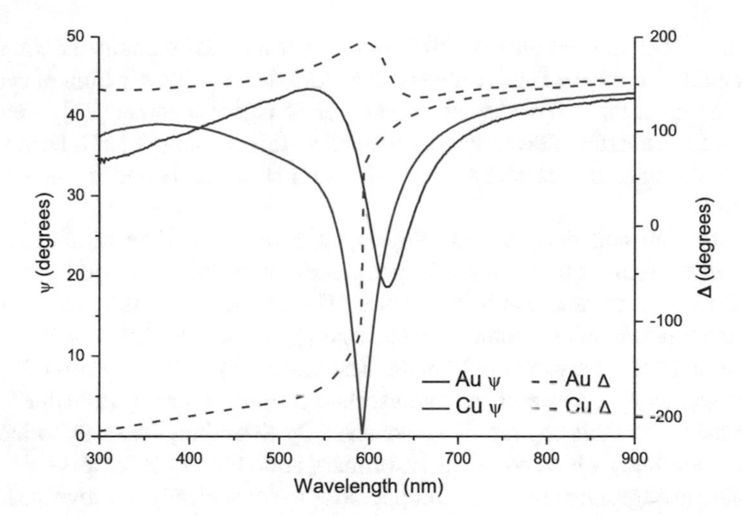

Fig. 7.2 Ellipsometric parameters ψ (solid curves) and Δ (dashed curves) showing surface plasmon resonances in air for a gold film (thickness 47.5 nm, AOI 47°, maroon curves) and copper thin film (thickness 43.5 nm, AOI 46.5°, blue curves) excited in the Kretschmann configuration. The more Fano-like profile of the surface plasmon resonance in gold compared to copper is due to the stronger interband absorption in gold at higher wavelengths

In addition to possessing narrower, higher-quality resonances than gold thin films, graphene-protected copper and silver surface plasmon resonances can exhibit resonances with minima close to or exactly touching the point of zero reflection. These points of zero reflection are also known as points of topological darkness and give rise to large jumps in the phase of reflected light (described below) [11, 12]. It has been shown that, in principle, such phase jumps can act as the basis for biosensing with a sensitivity 10^4 times that of commercially-available SPR systems [13].

Here we demonstrate the use of graphene-protected copper for amplitude- and phase-sensitive detection of HT-2 mycotoxin. HT-2 mycotoxin (mass $\sim$ 424 Da) is the main metabolite of T-2 mycotoxin [14]. Both mycotoxins are created by moulds that grow on improperly stored grains such as barley and wheat [15, 16]. A small number of major outbreaks of illnesses resulting from HT-2 and T-2 mycotoxin contamination of grains have been recorded in the 20th century. In general the effects of exposure (including vomiting, diarrhoea, skin irritation and bleeding) are acute, but high exposures can lead to the development of alimentary toxic aleukia, which can result in death [17]. T-2 mycotoxin contamination of grain stores in the Orenburg Oblast in the 1940s resulted in the deaths of over a hundred thousand people [18]. HT-2 and T-2 mycotoxins can also be absorbed through the skin; it has been alleged that "yellow rain", a chemical weapon used in Vietnam and Laos in the 1970s, contained T-2 mycotoxin [19].

A 2015 study [20] of 154 beer samples commonly available in Europe showed that HT-2 mycotoxin (which remains present throughout the beer brewing processes [21]) was present in 9.1% of studied samples. The contaminated samples contained an average HT-2 concentration of 30.9 ng/mL. Current EU recommendations state that screening techniques for T-2 and HT-2 mycotoxins must have a limit of detection (LOD) no higher than 10 μg/kg individually or 25 μg/kg together [22]. It has been shown that commercial SPR systems can achieve this sensitivity [23]; however, the immunoassay used in this study was unable to distinguish between T-2 and HT-2 mycotoxins.

Recently, our collaborators have developed an immunoassay which is specific to HT-2 mycotoxin [24]. A relatively poor LOD of 10 ng/mL (~500 μg/kg) was achieved using a commercial SPR system. (The specificity of the assay was confirmed using neosolaniol, another tricothecene mycotoxin very similar in chemical composition to HT-2 mycotoxin.) The highest sensitivity of 0.38 ng/mL (19 μg/kg) was realised using time-resolved fluorescence resonance energy transfer (FRET) measurements, but this could only be achieved by extending the assay to include a secondary antibody which bound to the primary antibody-HT-2 complex.

We have used graphene-protected copper SPR to dramatically improve the LOD of this specific HT-2 immunoassay. We have shown that measurements of ψ (amplitude of reflected light) allow for the detection of concentrations of HT-2 mycotoxin as low as 1 pg/mL, an improvement of a factor of 10^4 over the sentivity achieved with commercial SPR systems. Measurements of phase allow for a further enhancement of sensitivity of approximately a factor of 7. We have also shown that this technique can be used to detect HT-2 mycotoxin in beer. Our results show that graphene-protected copper can be used for ultra-sensitive detection in important, real-world applications.

7.2 Phase Sensitivity at Points of Topological Darkness

Let us consider the reflection of light on a thin film of fixed thickness at a certain angle of incidence with a refractive index $\hat{n} = n + ik$. For such a film there exists a set of points $(n(\lambda), k(\lambda))$ which give a reflection amplitude of zero. If one of these points of zero reflection coincides with the value of $\hat{n}$ for a real material of the same thickness at the same angle and wavelength then this corresponds to a point of topological darkness.

It has previously been shown that gold nanoparticle arrays [13] and self-assembled silver nanoparticle arrays [25] possess such points of topological darkness. The theoretical study of topological darkness in thin copper films is ongoing but the amplitude and phase behaviour of the copper thin film plotted in Fig. 7.2 suggests that for reasons outlined below it is very likely that copper also possesses points of topologically protected darkness.

When the amplitude of a reflected electromagnetic wave is zero its phase is undefined. As the system crosses a point of zero reflection it undergoes a π jump in the phase [11]. Small changes in reflection close to points of zero reflection can result in extremely high phase sensitivity.

Topological darkness does not only appear in plasmonic systems. Points of zero reflection also occur at the Brewster angle for p-polarised light, for example.

However, plasmonic systems remain the only ones capable of providing topological darkness alongside strong field enhancements, which has sustained their study for biosensing applications.

7.3 Sample Preparation

7.3.1 Graphene-Protected Copper Plasmonic Films

Silver thin films possess slightly higher quality plasmon resonances than copper thin films. However, silver reacts with water much more vigorously than copper. Therefore, the only graphene transfer method compatible with silver is the dry transfer of mechanically exfoliated graphene. This limits both the areas of graphene-protected silver and the potential scalability of graphene-protected silver. In contrast, copper thin films undergo relatively little oxidation when briefly exposed to water as in the wet transfer process for CVD graphene. Therefore, graphene-protected copper was used.

Copper thin films were deposited on glass substrates (thickness 1 mm, lateral dimensions 25 mm × 25 mm) using electron-beam lithography (Sect. 2.2.4). In principle one could grow CVD graphene directly on to the surface of the copper; in reality the CVD process roughens the copper surface and destroys the plasmonic properties of the film.

The growth and transfer of CVD graphene largely followed the standard procedures outlined in Sect. 3.4.2. Our CVD graphene was provided by BGT Materials and Graphenea. The transfer of graphene on to copper thin films presented a number of challenges uncommon to most transfer procedures. Before transfers, substrates are usually cleaned using O_2/Ar plasma etching, but the reactivity of copper with oxygen forbade this. Instead, graphene was transferred on to copper thin films immediately after their deposition. To prevent damage to the copper film from any residual copper etchant left on the graphene the graphene was thoroughly washed in deionised water before its final transfer on to the copper film. Immediately after the final transfer of graphene on to the substrate the sample is usually dried with a nitrogen gun and then annealed on a hot plate to improve adhesion. However, doing so would oxidise the copper film. As an alternative the samples are annealed in an Ar/H_2 atmosphere in a furnace at 150 °C for three hours.

The adhesion and overall quality of the graphene transfer on to the copper was checked using optical microscopy after its initial drying, after annealing and finally after removal of the PMMA membrane. The plasmonic quality of films was verified in air and water using spectroscopic ellipsometry (Sect. 2.2.5).

7.3.2 Functionalisation of Graphene with HT-2 Antibodies

In general, surface functionalisation for biosensing experiments is completed in three steps, illustrated in Fig. 7.3. First, a linker molecule is attached to the surface. One end of the linker binds to the surface and the other end terminates with a functional group that will easily bond with the receptor. Second, receptors are bound to the end of the linkers. The receptor is usually either an antibody or an aptamer. Third, a blocker is introduced to react to any unbound sites on the linker.

Surface functionalisation in this study was performed by Henri Arola and Miika Soikkeli using an adapted version of Lien et al.'s carbon surface immobilisation method [26].

We used 1-Pyrenebutyric acid N-hydroxysuccinimide ester (1-PBA NHS) our as our linker (chemical structure shown in Fig. 7.3e). The pyrene end binds to graphene by $\pi - \pi$ stacking [27] and the NHS ester end can bind to the amide (–NH_2) group on our antibody. Our HT-2 mycotoxin antibody was HT2-10 Fab; the details of its synthesis are beyond the scope of this work and can be found in reference [24]. The blocker was ethanolamine.

The protocol for surface functionalisation of our graphene-protected copper chips is described in detail below and illustrated in Fig. 7.3:

1. Preparation of linker solution: 2 mg/mL 1-PBA NHS in 100% methanol (MeOH). Sonicate and then incubate for 1 h in room temperature without shaking to

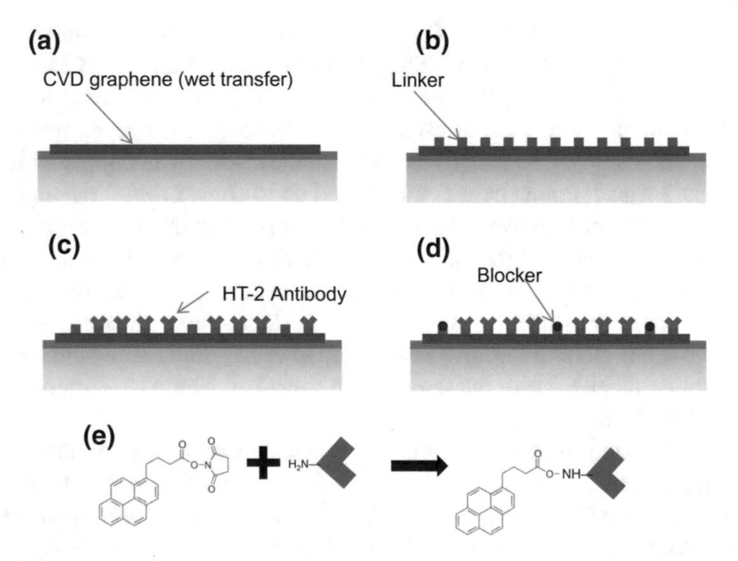

Fig. 7.3 Functionalisation of graphene surface. **a** Graphene-protected copper surface. **b** Binding of linkers (1-Pyrenebutyric acid N-hydroxysuccinimide ester) to graphene surface. **c** Binding of HT-2 mycotoxin antibodies to linkers. **d** Binding of blocker molecules to remaining unbound linker sites. **e** Binding mechanism of linker and HT-2 mycoxotin antibody

ensure saturation of solution. Remove undissolved linker from solution by filtering through a disposable filter unit attached to a syringe.

2. Immobilise linker on graphene surface (Fig. 7.3b): Incubate graphene-protected copper sample in 5 mL linker solution at room temperature for one hour. Wash sample in pure MeOH followed by PBS pH 7.3 (5 mL).
3. Bind antibodies to linkers (Fig. 7.3c,e): Incubate sample in 5 mL HT2-10 Fab in PBS pH 5 solution at a concentration of 50 μg/mL for 20 min at room temperature.
4. Add blockers to empty NHS linker sites (Fig. 7.3c): Incubate sample in 5 mL 100 mM ethanolamine solution (1 M ethanolamine stock solution (pH 8.5) diluted 1:10 in distilled water) for 10 min at room temperature.
5. Rinse chip in distilled water and store in distilled water before use.

The HT-2 mycotoxin solutions were prepared as follows. HT-2 mycotoxin was stored in a stock solution of dimethyl sulphoxide (DMSO, $(CH_3)_2SO$) at a concentration of 1 mg/mL. 10 μL HT-2 mycotoxin was added to 9.990 mL of 0.1 x phosphate buffered saline pH 7.3 to create 10 mL of 1000 ng/mL HT-2 mycotoxin solution. PBS solution is a standard solution of salts (usually chlorides and phosphates) that is used in biological experiments because it is isotonic with most cells. Studying cells in DI water would destroy the cells due to osmosis effects. Lower concentration solutions of HT-2 mycotoxin were prepared by diluting this initial solution with an appropriate volume of 0.1 x PBS containing 0.1% DMSO. The concentration of DMSO in the original solution was 0.1%; this DMSO concentration was maintained for lower concentrations of HT-2 mycotoxin for consistency.

7.4 Results

After sample preparation, we mounted the functionalised graphene-protected copper film in to the Kretschmann geometry. We first washed the sample with DI water and then with 0.1 M pH 7.4 phosphate-buffered saline solution (PBS solution). The solution was drawn through the cell (volume ~5 mL) using a mechanically controlled syringe.

Measurements were taken using spectroscopic ellipsometry (Sect. 2.2.5). The angle of incidence (typically 59–62°) was chosen to slightly detune from the point of zero reflection in order to provide sufficient light intensity at the minimum of the resonance for a reasonably noise-free reflection signal. When each new solution was drawn through the cell the system was left to stabilise for 10 min. We were unable to measure a dynamic scan of the binding of HT-2 mycotoxin to the antibodies on the graphene surface because we did not conduct the measurements in a temperature-controlled environment, meaning the thermally-induced shifts of plasmon resonances in our system were much larger than shifts produced by the HT-2 mycotoxin binding event. Ellipsometric reflection spectra were then taken at regular intervals for a further 5–10 min to ensure the stability of the system before introducing the next solution (with a concentration of HT-2 mycotoxin ten times the previously measured solution).

Figure 7.4 shows the typical shifting of the surface plasmon resonance measured in Ψ moving from toxin-free PBS to PBS solutions with increasingly high HT-2 mycotoxin concentration. The redshift of the surface plasmon resonance is clearly visible. If a solution of HT-2 mycotoxin was replaced with toxin-free PBS the spectral position of the plasmon resonance did not change. Therefore, the plotted redshifts can be attributed solely to the nonreversible binding events close to the surface of the graphene-protected copper.

This particular experiment utilised a secondary antibody (anti-IC-HT2-10, in all solutions at a concentration of 5 μg/mL) which binds to the conjugate of the primary antibody and the HT-2 toxin. In previous experimental studies the binding of the secondary antibody was necessary to detect the initial binding of HT-2 toxin to the antibody [24]; however, in our assay we found that the secondary antibody provided no enhancement of sensitivity compared to later experiments where no secondary antibody was used (see Fig. 7.7).

7.4.1 Volumetric Limit of Detection

Figure 7.5a shows the change in the ellipsometric parameter Ψ as a function of increasing toxin concentration in PBS. The change in Ψ from toxin-free PBS to 1 pg/mL HT-2 concentration is 1.22°, substantially higher than our measurement sensitivity of 0.05°. We have therefore shown that even for amplitude reflection

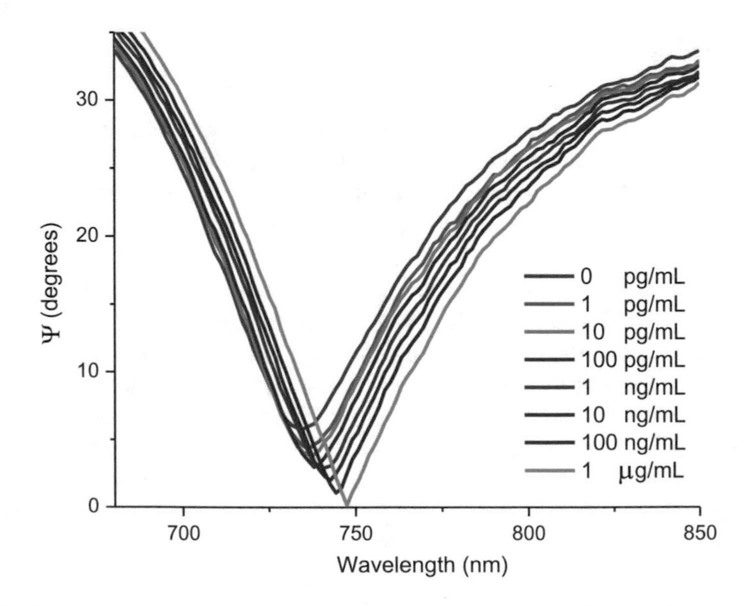

Fig. 7.4 Typical change in ellipsometric function Ψ moving from buffer solution to increasing concentrations of HT-2 mycotoxin. Angle of incidence 59°

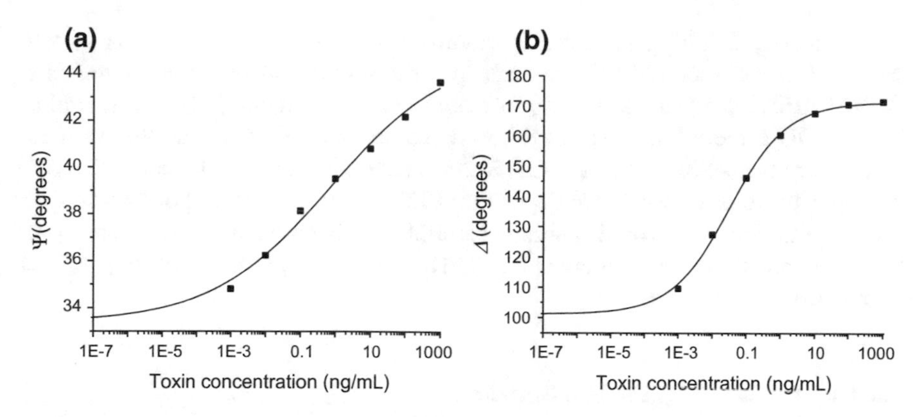

Fig. 7.5 Change in the ellipsometric parameters **a** Ψ (at $\lambda = 782$ nm) and **b** Δ ($\lambda = 922$ nm) with increasing HT-2 toxin concentration without the use of a secondary antibody. Angle of incidence 60°. Both curves have been fitted with Hill plots ($k_H = 0.80 \pm 0.72$ ng/mL, $n_H = 0.25 \pm 0.04$ for (**a**); $k_H = 0.033 \pm 0.004$ ng/mL, $n_H = 0.51 \pm 0.03$ for (**b**))

measurements we can easily detect HT-2 mycotoxin concentrations with at least four orders of magnitude greater sensitivity than measurements using this same assay in commercial gold-based SPR systems [24].

Figure 7.5b shows the change in the ellipsometric parameter Δ for the same sample as in Fig. 7.5a. One can see that the changes in phase due to increases in HT-2 concentration are over six times greater than the changes in amplitude reflection measurements. Here the introduction of 1 pg/mL HT-2 mycotoxin results in a change in Δ of 7.78°. Our system (with a sensitivity of 0.05°) therefore has a LOD of

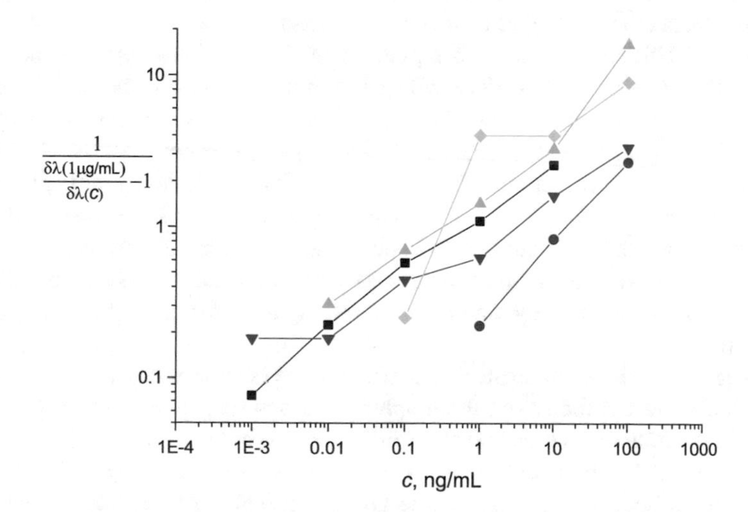

Fig. 7.6 Phase response to binding of HT-2 mycotoxin in beer ($\lambda = 841$ nm). Hill plot coefficients: $k_H = 0.35 \pm 4.60$ ng/mL and $n_H = 0.38 \pm 0.98$

approximately 6 fg/mL. A thermally stabilised advanced phase detection system could allow for a tenfold enhancement in the resolution of phase measurements, which in principle would further improve our LOD to 600 ag/mL [12]. This is similar to the LOD achieved in recent SPR-based studies [28–33]; however, these results rely either on labelling of the analyte [28] or on complicated sample fabrication (for example, using electrochemical deposition [32], self-assembly [33] or by covering the gold thin film with a of bis-aniline-crosslinked gold nanoparticle composite [30, 31]). In contrast, our analytes are unlabelled and the preparation of our samples is in comparison very simple.

7.4.1.1 Fitting Using the Hill Equation

The plots in Fig. 7.5 have been fitted to the Hill equation. The Hill equation was originally formulated to describe how the binding of ligands to receptor sites on proteins at equilibrium changes as a function of ligand concentration [34]:

$$f = \frac{1}{\left(\frac{k_H}{c}\right)^{n_H} + 1}. \tag{7.1}$$

f is the fraction of sites occupied by ligands, c is the volumetric concentration of ligands in solution, k_H is the ligand concentration required to occupy half of all available receptor sites, and n_H is the Hill coefficient. n_H describes the *cooperativity* of ligand binding: that is, how the binding affinity of sites on the surface changes as the ligands bind to sites. Positive cooperativity ($n_H > 1$) occurs when the binding of a ligand increases the binding affinity of the remaining sites on the protein. Negative cooperativity ($n_H < 1$) corresponds to the opposite case, where the binding of a ligand decreases the binding affinity of the remaining sites.

While the Hill equation describes binding of ligands to various sites on a protein, we are studying a 2D surface where individual binding sites (antibodies) do not interact. This would correspond to $n_H = 1$ and reduce to a Langmuir isotherm (initially formulated to describe the adsorption of gas molecules on a surface). However, Langmuir isotherms commonly provide poor fits to SPR data plots [35]. Indeed, the best fits for our SPR data curves consistently have $n_H < 1$, suggesting that the average binding affinity of antibodies in our assay decreases as the concentration of HT-2 mycotoxins increases. Since there is no way for the antibodies on the surface to interact we do not expect any kind of cooperativity as in the case of ligand receptor sites on proteins.

Instead we attribute our small Hill coefficients to heterogeneity in the binding of the antibodies to the linkers on the graphene surface (and potentially also heterogeneity in the antibodies themselves). Antibodies are bound to the surface when an amide (NH_2) group binds to the NHS-ester group on the linkers (Fig. 7.3e). However, since antibodies have many amide groups on their surface, the orientation of the antibodies with respect to the surface will be non-uniform. This means that some binding sites will be more easily accessible to HT-2 mycotoxins in solution than others, acting to change the effective binding affinity of antibodies. In this case we expect the antibodies with the highest binding affinities to bind to HT-2 mycotoxins first. Therefore, the average effective binding affinity of the vacant sites will decrease as the experiment progresses, giving an effect equivalent to negative cooperativity.

Here we recast the Hill equation in the context of analytes binding to antibodies on the surface of an SPR chip. It is not possible for us to directly measure f; instead we measure the change in Ψ, Δ or the position of the plasmon resonance.

Suppose that at the start of the sensing experiment, when the functionalised graphene-protected copper surface is in PBS but before any HT-2 mycotoxins have been introduced, the plasmon resonance occurs at a wavelength λ_i. We then introduce a volumetric concentration c of HT-2 mycotoxins into the solution. The HT-2 mycotoxins will bind to the antibodies on the surface. After a sufficiently long period of time (typically under 20 min) the system will have reached equilibrium and the plasmon resonance will have undergone a redshift $\delta\lambda(c)$. After repeating this process with progressively higher mycotoxin concentrations all antibodies will eventually bind to a mycotoxin. At this point the plasmon resonance will be at a wavelength λ_f.

If we assume the majority of antibodies are monoclonal (i.e. each antibody only has one binding site), then we expect there to be a linear dependence between f and $\delta\lambda(c)$:

$$f(c) = \frac{\delta\lambda(c)}{\lambda_f - \lambda_i}. \tag{7.2}$$

Equivalent relationships between f and Ψ or Δ can easily be derived. Combining Eqs. 7.1 and 7.2 yields the following relation[1]:

$$\frac{1}{(\lambda_f - \lambda_i)/\delta\lambda - 1} = \left(\frac{c}{k_H}\right)^{n_H}. \tag{7.3}$$

When plotted as a log-log plot this yields a straight line with gradient n_H and y-intercept $n \ln k_H$. Figure 7.7 plots this relationship for all experimental data collected. λ_f is not an experimentally known concentration (it corresponds to an asymptote). In this plot we have made the approximation of $\lambda_f \approx \lambda_i + \delta\lambda(1\ \mu\text{g/mL})$ since preliminary fitting suggests the difference between these values is small compared to $\lambda_f - \lambda_i$. In general, the values of n_H are very consistent, irrespective of either the presence of the secondary antibody or the solution used (PBS or beer), suggesting this approximation is a reasonable one.

[1]This equation is sometimes referred to as Philip's First Law, in honour of its discoverer.

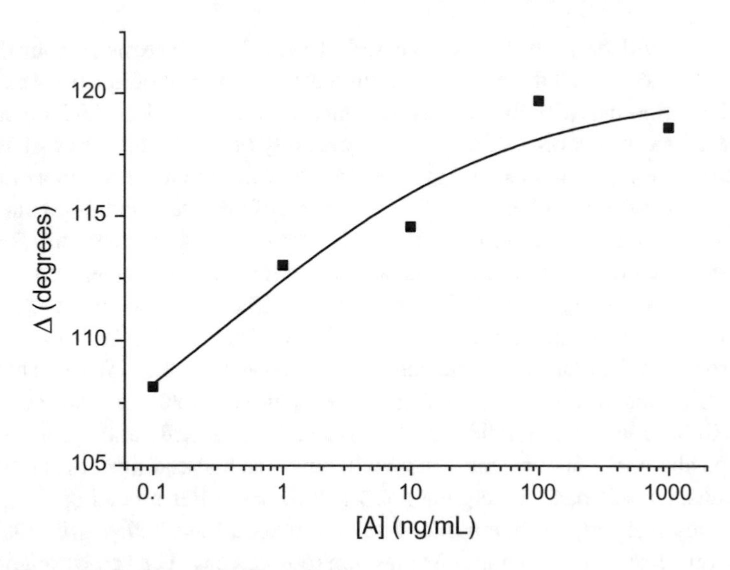

Fig. 7.7 Plot of Eq. 7.3 for all data. Red, black and green points correspond to experiments done in PBS without the use of the secondary antibody; blue points were measured in PBS with the secondary antibody; cyan points were measured in beer without secondary antibodies. Taking the weighted mean of the linear fits of these curves gives $k_H = 0.85 \pm 0.06$ ng/mL and $n_H = 0.35 \pm 0.02$

7.4.1.2 Measurements in Beer

Since one of the main applications is in food safety (especially beer) we have shown that our assay works not just in PBS but also in beer. Prior to SPR experiments, the beer samples were degassed overnight at room temperature in a container with a wide mouth. Shaking the sample before being left overnight and before experiments improved the removal of bubbles. We measured the shifts in plasmon caused by bottled and canned version of a well-known lager[2] relative to PBS. When returning from beer to PBS and measuring the spectral position of the plasmon resonance we observed no overall redshifts which can be attributed to the binding of HT-2 mycotoxin. In fact, more blueshifts were measured than redshifts. Since blueshifts of the plasmon resonance are due to temperature effects (cooler liquids have higher densities and so higher refractive indices) and small ($|\Delta\lambda| < 5$ nm), we conclude that any binding effects on this timescale are small compared to temperature effects. This suggests that if any HT-2 mycotoxin is present in solution it is present at much lower concentrations than those we have studied above.

To show that our assay is capable of detecting HT-2 mycotoxin in beer, we spiked beer with HT-2 mycotoxins with concentrations ranging from 0.1 ng/mL up to 1 μg/mL. This concentration range was chosen because this reflects the natural contamination range of HT-2 mycotoxin in beer [20, 36]. The preparation of solutions

[2]Stella Artois, 4.8% ethanol.

for this experiment was the same as described above for PBS-based measurements but with degassed beer used instead of 0.1 x PBS pH 7.

Phase-sensitive detection of HT-2 mycotoxin in beer is shown in Fig. 7.6. The sensitivity of these measurements is substantially lower than those measured in PBS; nevertheless, we still have a limit of detection of approximately 1 pg/mL in beer, comfortably below the 500 pg/mL detection limit recommended by the European Union. These measurements show that our assay is capable of detecting HT-2 mycotoxin in beer with sensitivities that make the technique practical for use in food safety applications.

7.4.2 Areal Mass Sensitivity

An important figure of merit for plasmonic biosensors is their areal mass sensitivity. Figure 7.8 is an AFM image of graphene on platinum functionalised with the surface chemistry used in this study. It clearly shows the pins that correspond to linker/receptor sites. Based on this AFM image we estimate the surface density of

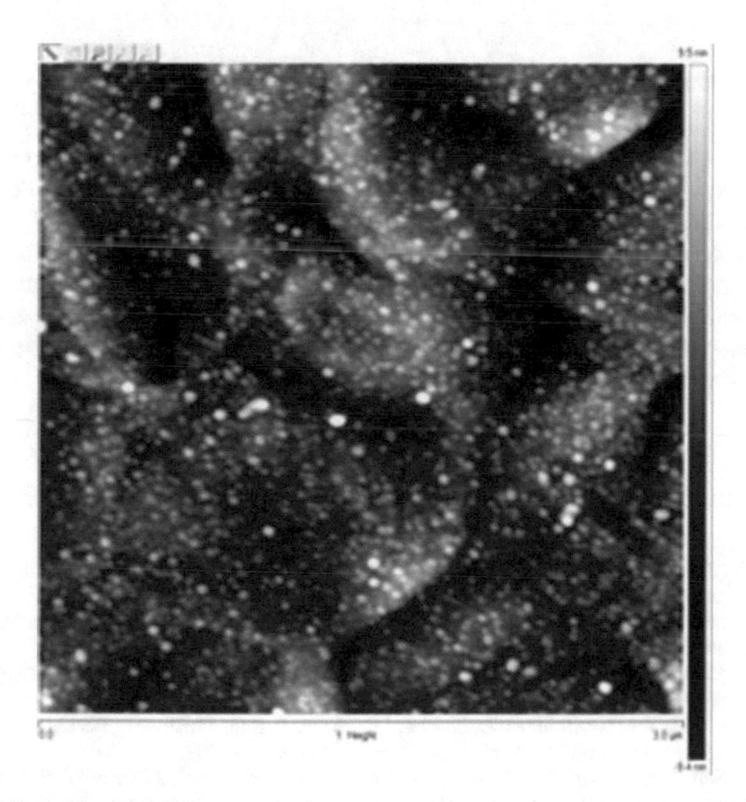

Fig. 7.8 AFM image of a graphene surface (on platinum instead of copper) functionalised with the surface chemistry used in this study

antibodies on the functionalised graphene to be no more than $100 \times 100\ \text{nm}^2$. However, since many of the linkers will have blockers at their ends instead of antibodies the surface density of antibodies could be much lower.

The areal mass sensitivity of our assay can be calculated using the surface density of antibodies, the mass of HT-2 mycotoxin (424 Da) and the initial and final values of the phase $\Delta_{i,f}$. We assume that the antibodies are predominantly monoclonal and that all available receptor sites have been occupied at Δ_f. The focused spot size is approximately $30 \times 60\ \mu\text{m}^2$ for an angle of incidence of 60°. If $\Delta_f - \Delta_i = 69.84°$ (Fig. 7.5) and the resolution is 0.05° then this gives an areal mass sensitive no higher than 50 ag/mm^2 (and below 5 ag/mm^2 after thermal stabilisation and advanced phase detection [12]). This is 4–5 orders of magnitude more sensitive than commercially available SPR systems [2] and suggests that we can resolve no more than a few tens of HT-2 mycotoxin molecules. We stress that this is an upper-bound estimate that assumes all linkers have antibodies attached; the actual areal mass sensitivity of our assay could be much lower. It is not possible to accurately estimate the proportion of linkers with antibodies from the AFM; this will more accurately be determined using quartz microbalance measurements currently being analysed by Henri Arola and Miika Soikkeli.

7.5 Conclusion

Graphene-protected copper plasmonics has been shown to be a highly competitive platform for SPR biosensing. We have demonstrated its sensitivity by detecting concentrations of HT-2 mycotoxins with four orders of magnitude greater sensitivity than commercial SPR systems are capable of. Phase-sensitive biosensing has shown to be particularly promising, with an areal mass sensitivity of just a few tens of attograms per square millimetre. Our results demonstrate that phase-sensitive graphene-protected copper plasmonic biosensing is a highly sensitive, practical, relatively simple biosensing technique with exceptional promise.

References

1. B. Liedberg, C. Nylander, I. Lundström, Biosensing with surface plasmon resonance—how it all started. Biosens. Bioelectron. **10**(8), i–ix (1995)
2. J. Homola, S.S. Yee, G. Gauglitz, Surface plasmon resonance sensors: review. Sens. Actuators B Chem. **54**(1), 3–15 (1999)
3. J.N. Anker, W.P. Hall, O. Lyandres, N.C. Shah, J. Zhao, R.P. Van Duyne, Biosensing with plasmonic nanosensors. Nat. Mater. **7**(6), 442–453 (2008)
4. X. Fan, I.M. White, S.I. Shopova, H. Zhu, J.D. Suter, Y. Sun, Sensitive optical biosensors for unlabeled targets: a review. Anal. Chim. Acta **620**(1), 8–26 (2008)
5. A.V. Kabashin, P. Evans, S. Pastkovsky, W. Hendren, G.A. Wurtz, R. Atkinson, R. Pollard, V.A. Podolskiy, A.V. Zayats, Plasmonic nanorod metamaterials for biosensing. Nat. Mater. **8**(11), 867–871 (2009)

6. V.G. Kravets, F. Schedin, A.V. Kabashin, A.N. Grigorenko, Sensitivity of collective plasmon modes of gold nanoresonators to local environment. Opt. Lett. **35**(7), 956–958 (2010)
7. B.D. Thackray, V.G. Kravets, F. Schedin, G. Auton, P.A. Thomas, A.N. Grigorenko, Narrow collective plasmon resonances in nanostructure arrays observed at normal light incidence for simplified sensing in asymmetric air and water environments. ACS Photonics **1**(11), 1116–1126 (2014)
8. J.H.T. Luong, K.B. Male, J.D. Glennon, Biosensor technology: technology push versus market pull. Biotechnol. Adv. **26**(5), 492–500 (2008)
9. P.R. West, S. Ishii, G.V. Naik, N.K. Emani, V.M. Shalaev, A. Boltasseva, Searching for better plasmonic materials. Laser Photonics Rev. **4**(6), 795–808 (2010)
10. V.G. Kravets, R. Jalil, Y.-J. Kim, D. Ansell, D.E. Aznakayeva, B. Thackray, L. Britnell, B.D. Belle, F. Withers, I.P. Radko, Z. Han, S.I. Bozhevolnyi, K.S. Novoselov, A.K. Geim, A.N. Grigorenko, Graphene-protected copper and silver plasmonics. Sci. Rep. **4** (2014)
11. A.N. Grigorenko, P.I. Nikitin, A.V. Kabashin, Phase jumps and interferometric surface plasmon resonance imaging. Appl. Phys. Lett. **75**(25), 3917–3919 (1999)
12. A.V. Kabashin, S. Patskovsky, A.N. Grigorenko, Phase and amplitude sensitivities in surface plasmon resonance bio and chemical sensing. Opt. Exp. **17**(23), 21191–21204 (2009)
13. V.G. Kravets, F. Schedin, R. Jalil, L. Britnell, R.V. Gorbachev, D. Ansell, B. Thackray, K.S. Novoselov, A.K. Geim, A.V. Kabashin, A.N. Grigorenko, Singular phase nano-optics in plasmonic metamaterials for label-free single-molecule detection. Nat. Mater. **12**(4), 304–309 (2013)
14. K. Kuca, V. Dohnal, A. Jezkova, D. Jun, Metabolic pathways of T-2 toxin. Curr. Drug Metab. **9**(1), 77–82 (2008)
15. EFSA Panel on Contaminants in the Food Chain (CONTAM). Scientific opinion on the risks for animal and public health related to the presence of T-2 and HT-2 toxin in food and feed. EFSA J. **9**, 2481 (2011)
16. R. Krska, A. Malachova, F. Berthiller, H.P. Van Egmond, Determination of T-2 and HT-2 toxins in food and feed: an update. World Mycotoxin J. **7**(2), 131–142 (2014)
17. P.S. Steyn, Mycotoxins, general view, chemistry and structure. Toxicol Lett. **82**, 843–851 (1995)
18. N.A. Foroud, F. Eudes, Trichothecenes in cereal grains. Int. J. Mol. Sci. **10**(1), 147–173 (2009)
19. S.A. Watson, C.J. Mirocha, A.W. Hayes, Analysis for trichothecenes in samples from southeast asia associated with "yellow rain". Fundam. Appl. Toxicol. **4**(5), 700–717 (1984)
20. Y. Rodríguez-Carrasco, M. Fattore, S. Albrizio, H. Berrada, J. Mañes, Occurrence of fusarium mycotoxins and their dietary intake through beer consumption by the european population. Food Chem. **178**, 149–155 (2015)
21. T. Inoue, Y. Nagatomi, A. Uyama, N. Mochizuki, Fate of mycotoxins during beer brewing and fermentation. Biosci. Biotechnol. Biochem. **77**(7), 1410–1415 (2013)
22. European Commission. EU: commission Recommendation of 27 March 2013 on the presence of T-2 and HT-2 toxin in cereals and cereal products. Off. J. Eur. Un. **165**(OJ L 91), 12–15 (2013)
23. J.P. Meneely, M. Sulyok, S. Baumgartner, R. Krska, C.T. Elliott, A rapid optical immunoassay for the screening of T-2 and HT-2 toxin in cereals and maize-based baby food. Talanta **81**(1), 630–636 (2010)
24. H.O. Arola, A. Tullila, H. Kiljunen, K. Campbell, H. Siitari, T.K. Nevanen, Specific noncompetitive immunoassay for HT-2 mycotoxin detection. Anal. Chem. **88**(4), 2446–2452 (2016)
25. L. Malassis, P. Massé, M. Tréguer-Delapierre, S. Mornet, P. Weisbecker, P. Barois, C.R. Simovski, V.G. Kravets, A.N. Grigorenko, Topological darkness in self-assembled plasmonic metamaterials. Adv. Mater. **26**(2), 324–330 (2014)
26. T.T.N. Lien, X.V. Nguyen, M. Chikae, Y. Ukita, Y. Takamura, Development of label-free impedimetric hCG-immunosensor using screen-printed electrode. J. Biosens. Bioelectron. **2**, 3 (2011)
27. V. Georgakilas, M. Otyepka, A.B. Bourlinos, V. Chandra, N. Kim, K.C. Kemp, P. Hobza, R. Zboril, K.S. Kim, Functionalization of graphene: covalent and non-covalent approaches, derivatives and applications. Chem. Rev. **112**(11), 6156–6214 (2012)

28. S. Krishnan, V. Mani, D. Wasalathanthri, C.V. Kumar, J.F. Rusling, Attomolar detection of a cancer biomarker protein in serum by surface plasmon resonance using superparamagnetic particle labels. Angew. Chem. Int. Ed. **50**(5), 1175–1178 (2011)
29. H.R. Sim, A.W. Wark, H.J. Lee, Attomolar detection of protein biomarkers using biofunctionalized gold nanorods with surface plasmon resonance. Analyst **135**(10), 2528–2532 (2010)
30. M. Riskin, R. Tel-Vered, O. Lioubashevski, I. Willner, Ultrasensitive surface plasmon resonance detection of trinitrotoluene by a bis-aniline-cross-linked au nanoparticles composite. J. Am. Chem. Soc. **131**(21), 7368–7378 (2009)
31. Y. Ben-Amram, R. Tel-Vered, M. Riskin, Z.-G. Wang, I. Willner, Ultrasensitive and selective detection of alkaline-earth metal ions using ion-imprinted Au NPs composites and surface plasmon resonance spectroscopy. Chem. Sci. **3**(1), 162–167 (2012)
32. P.L. Truong, C. Cao, S. Park, M. Kim, S.J. Sim, A new method for non-labeling attomolar detection of diseases based on an individual gold nanorod immunosensor. Lab Chip **11**(15), 2591–2597 (2011)
33. J.-H. Lee, B.-C. Kim, B.-K. Oh, J.-W. Choi, Highly sensitive localized surface plasmon resonance immunosensor for label-free detection of HIV-1. Nanomed. Nanotechnol. Biol. Med. **9**(7), 1018–1026 (2013)
34. D.L. Nelson, M.M. Cox, *Lehninger Principles of Biochemistry*, 6th edn. (W. H. Freeman, 2013)
35. P. Schuck, H. Zhao, *The Role of Mass Transport Limitation and Surface Heterogeneity in the Biophysical Characterization of Macromolecular Binding Processes by SPR Biosensing* (Humana Press, Totowa, NJ, 2010), pp. 15–54
36. J. Rupert, C. Solar, R. Marín, K.L. James, J. Mañes, Mass spectrometry strategies for mycotoxins analysis in European beers. Food Control **30**, 122–128 (2013)

Chapter 8
Conclusions and Future Work

In this thesis we have investigated two different methods by which plasmon resonances can be narrowed and demonstrated their feasibility in a number of applications. Nanostripe arrays fabricated on a gold sublayer have been shown to give extremely high-quality plasmon resonances at telecommunication wavelengths and their application demonstrated in novel mechanical electro-optical modulator and waveguide designs. Meanwhile, graphene- protected copper thin films have been shown to allow plasmonic biosensing with a sensitivity at least 4 orders of magnitude greater than what is attainable using commercial gold film-based biosensors.

Diffraction coupling of localised plasmon resonances in nanoparticle arrays provides a way of dramatically enhancing the quality factor of plasmon resonances. In Chap. 4 we showed that diffraction coupling of gold nanostripe arrays on a metallic sublayer can give rise to extremely narrow plasmon resonances at 1.5 μm. For the first time we showed it is possible to achieve quality factors of 300 at 1.5 μm and studied the properties of these resonances as a function of nanostripe height. Two optoelectronic proof-of-concept devices utilising this nanoarray design were demonstrated in Chaps. 5 and 6. The sharpness of these resonances could make them ideal for plasmonic biosensing experiments, something that could form the basis of future studies.

In Chap. 5 a novel design for a nanomechanical electro-optical modulator was demonstrated. The modulator consists of the nanostripe array, a graphene/hBN heterostructure and an air gap that naturally forms between these two components due to the inherent roughness of the nanostripe array. Applying a gate voltage across this device causes the graphene/hBN stack to move closer to the nanostripe array, resulting in strong modulation effects from the ultraviolet through to the mid-infrared. Modelling of the device suggests that the magnitude of this modulation effect could be further enhanced by reducing the size of the air gap from 200–300 nm

P. A. Thomas, *Narrow Plasmon Resonances in Hybrid Systems*,
Springer Theses, https://doi.org/10.1007/978-3-319-97526-9_8

to below 100 nm. Attempts to fabricate such a modulator (using a mirror instead of a plasmonic nanoarray) have been led by Francisco Rodriguez.

The nanostripe array was used as the basis for a new hybrid plasmonic waveguide design. Strong coupling was achieved between the diffraction coupled plasmon resonances in the nanostripe array and waveguide modes in a 400 nm-thick layer of HfO_2 deposited on top of the array. It was demonstrated that light could be guided over a transmission length of approximately 250 μm. Two outstanding issues require further study. First, although we demonstrated a communication length of 250 μm, we did not measure the propagation length of hybrid plasmon-waveguide modes in this structure, the standard figure of merit for plasmonic waveguides. To measure this figure we would need to fabricate waveguides of a variety of lengths and measure how the intensity of the light decoupled from each structure varies as a function of waveguide length. Additionally, further work is required to understand the origin of the anomalous resonances for s-polarised guided modes shown in Fig. 6.5b, d.

Graphene-protected copper thin films have already been shown to provide substantially stronger, narrower plasmons resonances than gold thin films with high phase sensitivity. In Chap. 7 we have for the first time used graphene-protected copper as a platform for SPR biosensing. We have measured the binding of HT-2 mycotoxin on functionalised graphene-protected copper at concentrations as low as 1 pg/mL, an improvement of four orders of magnitude over what could be detected using the same HT-2 mycotoxin assay with commercial gold-based SPR. Phase-sensitive measurements provide at least a sixfold enhancement in sensitivity over amplitude-sensitive measurements. We estimate the volumetric limit of detection to be 6 fg/mL and the areal mass sensitivity to be no more than 50 ag/mm^2 (a few tens of molecules in one illuminated spot). These figures rank phase-sensitive graphene-protected copper biosensing among the most sensitive assays in the literature on next-generation SPR biosensing techniques. We also showed that our assay provides practical levels of sensitivity when detecting HT-2 mycotoxin in beer. Previous studies of the HT-2 mycotoxin assay have already shown that the antibody used is specific to HT-2 mycotoxin and insensitive to any other similar mycotoxins. Nevertheless, future studies should verify that this remains the case when the antibody is used in a graphene-protected copper assay. In principle, it is also possible to improve the sensitivity of our technique further while retaining specificity by replacing the relatively bulky antibodies used in our study with smaller aptamers. An aptamer-based assay is currently being developed by Philip Day with experimental tests being carried out by Fan Wu.

Printed in the United States
By Bookmasters